Lecture Guide

to accompany

Beginning and Intermediate Algebra:

The Language and Symbolism of Mathematics

Third Edition

James W. Hall
Parkland College

Brian A. Mercer
Parkland College

Prepared by
Kelly Bails
Parkland University

The McGraw-Hill Companies

Lecture Guide for use with
BEGINNING AND INTERMEDIATE ALGEBRA: THE LANGUAGE AND SYMBOLISM
OF MATHEMATICS, THIRD EDITION
JAMES W. HALL AND BRIAN A. MERCER

Published by McGraw-Hill Higher Education, an imprint of The McGraw-Hill Companies, Inc., 1221 Avenue of the Americas, New York, NY 10020.

This book is printed on recycled, acid-free paper containing 10% post consumer waste.

2 3 4 5 6 7 QVS/QVS 20 19 18 17 16

ISBN: 978-0-07-729691-9
MHID: 0-07-729691-5

www.mhhe.com

CONTENTS

PREFACE

Lecture Guide for Beginning and Intermediate Algebra: The Language and Symbolism of Mathematics

To the student:

Each section of this *Lecture Guide* is outlined like a class lecture with key information missing. Many of the items that could consume lots of your time recopying are provided for you. What is missing is a key word or an important step which requires your thinking and active mental participation. By copying less, you can concentrate and think more.

If your instructor uses this *Lecture Guide* as a framework for day-by-day class activities, you can follow along carefully, complete this *Lecture Guide*, and produce a well-organized set of class notes. You will have key definitions, key concepts, and a selection of examples and exercises. This process will also help improve your organization and note-taking skills for other classes.

You also can use this *Lecture Guide* for organizing your own review, for extra study, to structure your work outside of class, or as a guide to material you may have missed if you were not able to attend a class. It may be particularly useful to you if you are taking an online course.

If your class does not use the *Lecture Guide* as an in-class supplement and you have purchased it on your own for extra practice and review, ask your professor for access to a printout of the answers. The answers are available to your instructor on the Instructor Resource Site on the text's MathZone site for free through your school's McGraw-Hill campus sales representative.

Surveys of students using the *Lecture Guide* report an overwhelming feeling that the *Lecture Guide* helped the students succeed in their course and to learn how to learn mathematics.

To the instructor:

The *Lecture Guide* is designed to be used by novice and more experienced instructors and can serve as an outline for 80 to 95% of a lecture. Many instructors use the PowerPoint version of the *Lecture Guide* while others use transparencies and an overhead projector to present each lecture in the classroom.

The order of the examples and exercises follows the order of the objectives for each section of the textbook. The format is designed to keep each student involved completing the key information that is missing in the exercises. To do this, the *Lecture Guide* minimizes the amount of given information the student is required to recopy. For example, if an exercise requires a graph then a blank grid is provided. Then all students will have complete notes with graphs drawn to scale. If a table is needed, the given information will be included in the table with blank spaces for the new information to be inserted. For definitions and procedures, these are recopied in the *Lecture Guide* with key words or phrases omitted so they can be completed in class. The instructor and student can both focus on the key concept in this definition or procedure and then this key information can be emphasized by inserting it into the blanks that are provided. The *Lecture Guide* also includes questions that facilitate the development of multiple perspectives: algebraic, graphical, numerical, and verbal. One way this is done is by including questions requiring interpretation of given calculator graphs or tables or problems for calculator solution.

Results from student surveys:

Student surveys consistently yield positive feedback regarding the use of the *Lecture Guide* as an in-class supplement. Once completed, the *Lecture Guide* serves as an excellent set of notes for students to reference while working on homework and studying for exams. In surveys, students reported that they liked: spending less class time copying from the board and more time concentrating on the key ideas and writing only the more critical information, that their graphs and notes were more organized, and that is was easier to understand the purpose of each class. Faculty reported that their class evaluations improved and that student rating of the course increased. Department chairs report that using the *Lecture Guide* brought greater consistency to these courses, that students became more actively involved in the classes, and class retention increased.

History of the Lecture Guide:

Individual instructors have used versions of the *Lecture Guide* for many years. In 1999, data was collected at Parkland College comparing attrition rates and student success in all sections of Beginning Algebra. The favorable results of the sections using the *Lecture Guide* and the positive feedback from the students encouraged the department to recommend this supplement to all adjunct and full-time faculty. Ten years later this guide has been continuously improved and is still used in almost all sections of Beginning and Intermediate Algebra. Many of these faculty members have now created similar lectures guides for their other courses.

Lecture Guides for

Chapter 1

Operations with Real Numbers and a Review of Geometry

1.1 Lecture Guide: Preparing for an Algebra Class

Objective 1: Understand the class syllabus and textbook features.

Set Goals

1. List one academic goal that you plan to achieve in the first two weeks of class.

2. List one academic goal that you wish to achieve by the end of the semester.

Time Management

3. Complete the following chart by including class time, study time, sleep time, employment, or other time commitments. Remember to include 2 hours of study time for each hour of class.

	Sunday	Monday	Tuesday	Wednesday	Thursday	Friday	Saturday
12am-1:00							
1:00-2:00							
2:00-3:00							
3:00-4:00							
4:00-5:00							
5:00-6:00							
6:00-7:00							
7:00-8:00							
8:00-9:00							
9:00-10:00							
10:00-11:00							
11:00-Noon							
Noon-1:00							
1:00-2:00							
2:00-3:00							
3:00-4:00							
4:00-5:00							
5:00-6:00							
6:00-7:00							
7:00-8:00							
8:00-9:00							
9:00-10:00							
10:00-11:00							
11:00-12am							

Organize Your Work

4. Teachers often examine the ________________________ and ________________________ of a student's work to determine if the problems have been worked correctly. So, it is important to show all work in an organized format.

5. Always check the answers to the ________________________ numbered problems in the back of the text.

6. Always circle ________________________ answers on your homework, so you can ask about them in class.

7. Many teachers recommend putting all homework, quizzes and tests in a ______________________.

Getting the Most Out of the Lectures

8. To get the most out of the lectures, you should

(a) ____

(b) ____

(c) ____

Study in a Group

9. Find 3 or 4 classmates and exchange the following information. It would be a good idea to schedule a time for the group to meet to go over homework, ask/answer questions, or prepare for an exam.

Name	Phone Number	Email address

Calculator Usage

10. In addition to learning how to use the memory and parentheses keys, the two special keys you should use are __________ and __________. These two keys are accurate from two to three more digits than shown on the display.

Preparing for Tests

11. Practice tests are a good way to gauge your level of preparedness and help you overcome test anxiety. Three features in the text that you can use to prepare for exams are ________________________, ________________________, and ________________________.

12. What is another strategy that will help you prepare for tests?

1.2 Lecture Guide: The Real Number Line

Objective 1: Identify additive inverses.

Every real number has an additive inverse. This concept is important when we begin to look carefully at subtraction.

Opposites or Additive Inverses

Algebraically	**Verbally**	**Numerical Examples**	**Graphical Example**
If a is a real number, the *opposite* of a is _____.	Except for zero, the additive inverse of a real number is formed by changing the _______ of the number.	_____ is the opposite of 3 _____ is the opposite of -3 0 is the opposite of 0	Opposites -3 0 3
$a+(-a)=$ _____ $-a+a=$ _____	The sum of a real number and its additive inverse is ________.	$3+(-3)=0$ $-3+$ _____ $=0$ $0+0=0$	

Write the additive inverse of each number.

1. Number: -2

Additive Inverse: _______

2. Number: $\sqrt{5}$

Additive Inverse: _______

3. Number: $\frac{1}{3}$

Additive Inverse: _______

4. Number: 0

Additive Inverse: _______

One number is graphed on each of the following number lines. Graph the additive inverse of each number on the same number line.

5. 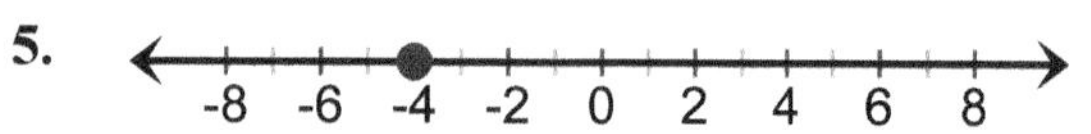

6. 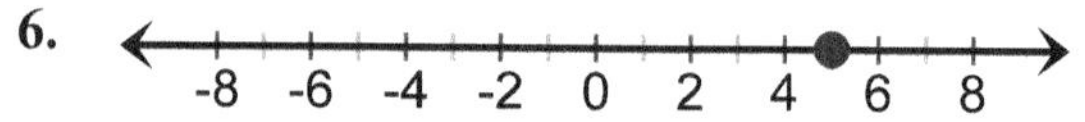

Double Negative Rule

Algebraically	**Verbally**	**Numerical Example**
For any real number a, $-(-a)=a$.	The opposite of the additive inverse of a is a.	$-(-7)=$ ____________

Simplify each expression.

7. $5.1+(-5.1)$

8. $6.4+0$

9. $-(-3)$

10. $-[-(-3)]$

Objective 2: Evaluate absolute value expressions.

Absolute Value

Algebraically	**Verbally**	**Numerical Example**	**Graphical Example**
$\lvert x\rvert = \begin{cases} x & \text{if } x \text{ is nonnegative} \\ -x & \text{if } x \text{ is negative} \end{cases}$	The absolute value of x is the ______________ between 0 and x on the number line.	$\lvert 2\rvert = 2$ $\lvert -2\rvert = 2$	2 units left 2 units right −2 0 2

For each number graphed below, determine the distance from this point to the origin, and give the absolute value of this number.

11. 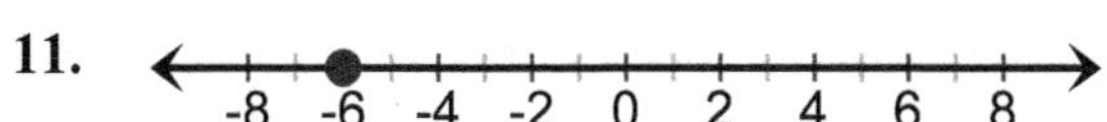

Distance: ______ Absolute value: ______

12. 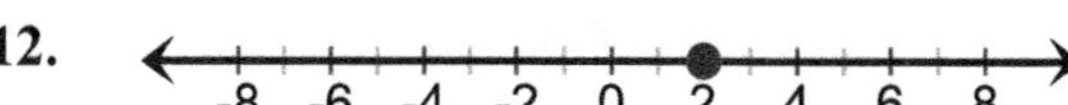

Distance: ______ Absolute value: ______

The absolute value of a nonzero number is always a positive value since distance is never negative.

Evaluate each absolute value expression.

13. $\lvert -8\rvert$ **14.** $\lvert 12\rvert$ **15.** $-\lvert 8\rvert$ **16.** $-\lvert -12\rvert$

17. If x is positive, the numerical value of the absolute value of x is **negative / zero / positive** (Circle the best choice.) and $\lvert x\rvert$ could be represented algebraically by **$-x$ / x** (Circle the best choice.).

18. If x is 0, the absolute value of x is ______.

19. If x is negative, the numerical value of the absolute value of x is **negative / zero / positive** (Circle the best choice) and $\lvert x\rvert$ could be represented algebraically by **$-x$ / x** (Circle the best choice).

20. Fill in the blanks to explain why the absolute value of x is defined in two parts.
Since distance is never negative, the absolute value of x requires a change in sign for values of x that are __________________ of zero on the number line and does not change the sign for values of x that are at zero or to the __________________ of zero on the number line.

Objective 3: Use inequality symbols and interval notation.

Equality and Inequality Symbols

Algebraic Notation	**Verbal Meaning**	**Graphical Relationship on the Number Line**
$x = y$	x equals y	x and y are the ____________ point.
$x \approx y$	x is approximately equal to y	x and y are close but are ____________ the same point.
$x \neq y$	x is not equal to y	x and y are ______________ points.
$x < y$	x is less than y	Point x is to the ____________ of point y.
$x \leq y$	x is less than or equal to y	Point x is on or to the ____________ of point y.
$x > y$	x is greater than y	Point x is to the ____________ of point y.
$x \geq y$	x is greater than or equal to y	Point x is on or to the ____________ of point y.

Insert <, =, or > in the blank to make each statement true.

21. -4 ____ -5 **22.** $\frac{3}{4}$ ____ $\frac{3}{5}$ **23.** $|-4|$ ____ $-|-4|$ **24.** $\sqrt{5}$ ____ $|5|$ **25.** $|-9|$ ____ $|9|$

Interval Notation

Inequality Notation	**Verbal Meaning**	**Graph**	**Interval Notation**
$x > a$	x is ____________ than a	a	(a, ∞)
$x \geq a$	x is greater than or ____________ to a	a	$[a, \infty)$
$x < a$	x is ____________ than a	a	$(-\infty, a)$
$x \leq a$	x is less than or ____________ to a	a	$(-\infty, a]$
$a < x < b$	x is ____________ than a and ____________ than b	a b	(a, b)
$a < x \leq b$	x is ____________ than a and ____________ than or equal to b	a b	$(a, b]$
$a \leq x < b$	x is ____________ than or equal to a and ____________ than b	a b	$[a, b)$
$a \leq x \leq b$	x is ____________ than or equal to a and ____________ than or equal to b	a b	$[a, b]$
$-\infty < x < \infty$	x is any ____________ number		$(-\infty, \infty)$

26. In interval notation a parenthesis means that an endpoint **is / is not** (Circle the best choice.) included in the interval. A bracket means that an endpoint **is / is not** (Circle the best choice.) included in the interval.

27. The table below contains four ways to refer to a set of real numbers. Complete the following table by filling in the missing two columns from each row. It really helps to understand a symbolic notation if you can say the verbal description to yourself.

Verbal Description	Inequality Notation	Number Line Graph	Interval Notation
	$x \le 4$		$(-\infty, 4]$
x is greater than three.		3	
	$-1 < x < 6$	−1 6	
x is greater than or equal to -5 and less than 2.			$[-5,2)$

Objective 4: Mentally estimate square roots and use a calculator to approximate square roots.

28. Complete the following table of common square roots. To estimate a square root of a number, it is extremely helpful to first think of a perfect square near that number.

Determine without a calculator the exact value to complete each equation.		Estimate the following square roots to the nearest integer and fill in the relationship between the square root and your estimate with either < or >.	Use a calculator or a spreadsheet to approximate the following square roots to the nearest hundredth.
$\sqrt{1} = 1$	$\sqrt{64} = 8$	$\sqrt{11}$	$\sqrt{11} \approx$
$\sqrt{4} = 2$	$\sqrt{81} =$	$\sqrt{48}$	$\sqrt{48} \approx$
$\sqrt{} = 3$	$\sqrt{100} = 10$	$\sqrt{50}$	$\sqrt{50} \approx$
$\sqrt{16} = 4$	$\sqrt{121} =$	$\sqrt{125}$	$\sqrt{125} \approx$
$\sqrt{25} = 5$	$\sqrt{} = 12$	$\sqrt{99}$	$\sqrt{99} \approx$
$\sqrt{} = 6$	$\sqrt{169} =$	$\sqrt{192}$	$\sqrt{192} \approx$
$\sqrt{49} =$	$\sqrt{196} =$	$\sqrt{170}$	$\sqrt{170} \approx$
	$\sqrt{} = 15$		

Objective 5: Identify natural numbers, whole numbers, integers, rational numbers, and irrational numbers.

29. Give the definitions of the integers, the whole numbers, and the natural numbers.

Natural Numbers:

Whole Numbers:

Integers:

All real numbers are either rational or irrational.

Rational and Irrational Numbers

Rational			
Algebraically	**Numerically**	**Numerical Examples**	**Verbal Examples**
A real number x is rational if $x = \frac{a}{b}$ for integers a and b, with $b \neq 0$.	In decimal form, a rational number is either a ______________ decimal or an infinite repeating decimal.	$\frac{1}{2} = 0.5$	$\frac{1}{2}$ in decimal form is a terminating decimal.
		$\frac{1}{3} = 0.333... = 0.\overline{3}$	$\frac{1}{3}$ in decimal form is a repeating decimal.
		$\frac{5}{33} = 0.151515... = 0.\overline{15}$	$\frac{5}{33}$ in decimal form is a repeating decimal.
Irrational			
Algebraically	**Numerically**	**Numerical Example**	**Verbal Examples**
A real number x is irrational if it cannot be written as $x = \frac{a}{b}$ for integers a and b.	In decimal form, an irrational number is an infinite non-____________ decimal.	$\sqrt{2} \approx 1.414213$	$\sqrt{2}$ cannot be written as a rational fraction – it is an infinite non-repeating decimal.
		$\pi \approx 3.141593$	π cannot be written as a rational fraction – it is an infinite non-repeating decimal.
		$0.1010010001...$	This irrational number does exhibit a pattern but it does not terminate and it does not repeat.

The following diagram may be helpful to visualize how the subsets of the real numbers are related.

The Real Numbers

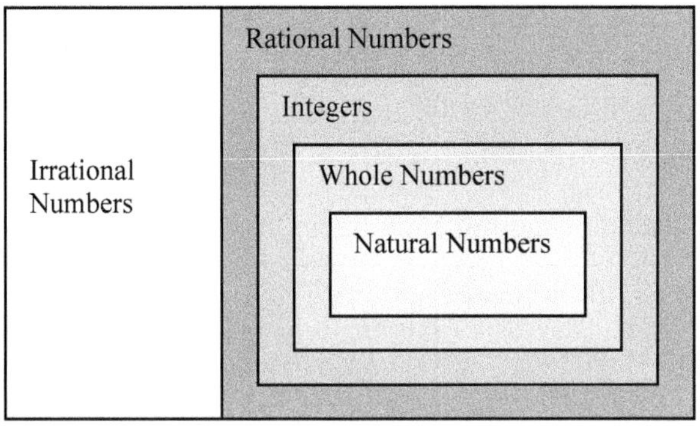

Irrational Numbers	Rational Numbers: Integers: Whole Numbers: Natural Numbers

30. Place a check mark in each column to which each number belongs.

Number	Natural	Whole	Integer	Rational	Irrational	Real
-3						
$\frac{2}{5}$						
π						
23						
$-2\frac{3}{4}$						
0						
$\sqrt{16}$						
$\sqrt{15}$						
$1.\overline{234}$						

($1.\overline{234}$ means 1.234234234...)

31. One column in problem 30 has a check mark for each number? Which column? ____________

32. Try evaluating $\sqrt{-4}$ and $\frac{3}{0}$ on a calculator or a spreadsheet. What happens?

33. Can you express the number 3 as a fraction and in decimal form? If so, provide an example.

34. Is the square root of 4 a rational number?

35. Is the square root of 5 a rational number?

1.3 Lecture Guide: Operations with Positive Fractions

Objective 1: Reduce fractions to lowest terms.

A positive fraction is in lowest terms if the numerator and the denominator are positive and have no common factor greater than _____. To reduce a fraction to lowest terms _______________ both the numerator and the ____________________ by their common _______________ .

1. Write a fraction in lowest terms to represent the shaded portion of each figure.

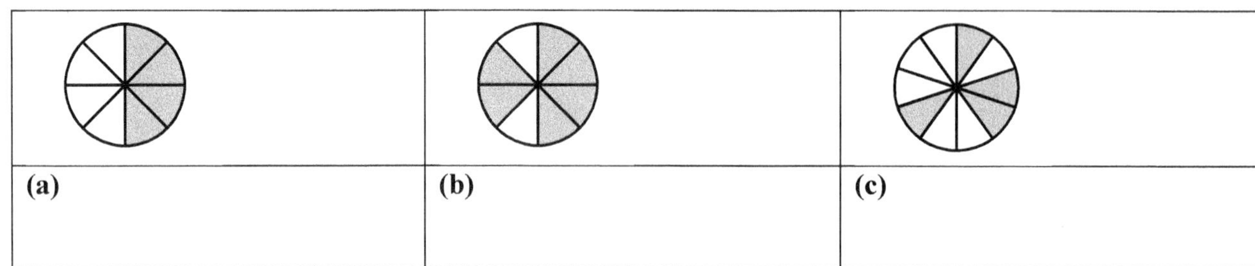

(a)	**(b)**	**(c)**

Reduce each fraction to lowest terms.

2. $\frac{24}{16}$ **3.** $\frac{15}{75}$ **4.** $\frac{36}{96}$ **5.** $\frac{22}{84}$

Objective 2: Multiply and divide fractions.

Multiplication of Fractions

Verbally	**Algebraically**	**Numerical Example**
To multiply two fractions, _______________ the numerators and multiply the denominators.	$\frac{a}{b}\cdot\frac{c}{d}=\frac{ac}{bd}$ for $b \neq 0$ and $d \neq 0$.	$\frac{2}{7}\cdot\frac{3}{5}=$

Perform the indicated multiplication and write the answer in lowest terms.

6. $\frac{2}{3}\cdot\frac{5}{7}$

7. $\frac{3}{4}\cdot\frac{6}{7}$

8. $\frac{5}{8}\cdot\frac{2}{15}$

9. $\frac{3}{5}\cdot 20$

10. $\frac{1}{2}\cdot\frac{10}{13}\cdot\frac{26}{35}$

11. Determine $\frac{3}{8}$ of 56.

Division of Fractions

Verbally	**Algebraically**	**Numerical Example**
To divide two fractions, multiply the first fraction by the ______________ of the second fraction.	$\frac{a}{b} \div \frac{c}{d} = \frac{a}{b} \cdot \frac{d}{c} = \frac{ad}{bc}$ for $b \neq 0$, $c \neq 0$, and $d \neq 0$.	$\frac{2}{3} \div \frac{5}{7} =$

12. (a) Why is it important that we require that $b \neq 0$, $c \neq 0$, and $d \neq 0$ in the rule for dividing fractions?

(b) Can integers like 4 be written as fractions?

Perform the indicated division and write the answer in lowest terms.

13. $\frac{2}{3} \div \frac{3}{5}$

14. $\frac{12}{5} \div \frac{4}{3}$

15. $5 \div \frac{3}{4}$

16. $\frac{3}{5} \div 4$

17. $\frac{7}{8} \div \frac{1}{2}$

18. Divide 56 by $\frac{7}{8}$.

Objective 3: Add and subtract fractions with the same denominator.

Addition and Subtraction of Fractions

Verbally	**Algebraically**	**Numerical Example**
To add fractions with the same denominator, add the _______________ and use the common denominator.	$\frac{a}{b}+\frac{c}{b}=\frac{a+c}{b}$ for $b \neq 0$.	$\frac{4}{7}+\frac{2}{7}=$
To add subtract fractions with the same denominator, subtract the _______________ and use the common denominator.	$\frac{a}{b}-\frac{c}{b}=\frac{a-c}{b}$ for $b \neq 0$.	$\frac{4}{7}-\frac{2}{7}=$

Perform the indicated additions and subtractions and express the result in lowest terms.

19. $\frac{3}{11}+\frac{4}{11}$

20. $\frac{1}{8}+\frac{3}{8}$

Perform the indicated additions and subtractions and express the result in lowest terms.

21. $\frac{8}{15}-\frac{3}{15}$

22. $\frac{11}{24}-\frac{7}{24}$

Objective 4: Add and subtract fractions with different denominators.

To add or subtract fractions with different denominators, we must first convert each fraction to an equivalent form having the _______________ denominator.

Write each fraction as an equivalent fraction with the given denominator.

23. $\frac{3}{5}=\frac{?}{20}$

24. $\frac{7}{12}=\frac{?}{36}$

Perform the indicated additions and subtractions and express the result in lowest terms.

25. $\frac{1}{5}+\frac{4}{15}$

26. $\frac{5}{8}-\frac{1}{2}$

27. $\frac{5}{12}+\frac{3}{8}$

28. $\frac{7}{15}-\frac{5}{18}$

29. $\frac{11}{20}+\frac{7}{30}$

30. $\frac{13}{16}-\frac{5}{24}$

Objective 5: Perform operations with mixed numbers.

Perform the indicated operations and express the result as a mixed number in lowest terms.

31. $6\frac{1}{2}+4\frac{1}{5}$

32. $6\frac{1}{2}-4\frac{1}{5}$

33. $5\frac{1}{4}\times 2\frac{2}{3}$

34. $5\frac{1}{4}\div 2\frac{2}{3}$

35. Adding Fractions vs. Multiplying Fractions: Add the fractions in the first column and multiply the fractions in the second column.

<u>Adding</u>

(a) $\frac{1}{12}+\frac{5}{12}$

(c) $\frac{2}{3}+\frac{1}{4}$

<u>Multiplying</u>

(b) $\frac{1}{12}\cdot\frac{5}{12}$

(d) $\frac{2}{3}\cdot\frac{1}{4}$

1.4 Lecture Guide: Addition and Subtraction of Real Numbers

Objective 1: Add positive and negative real numbers. There is a difference in the process for adding two numbers with like signs versus adding two numbers with unlike signs.

Adding Real Numbers with Like Signs

Verbally	Numerical Examples	Graphical Examples
To add numbers with like signs, first add the absolute values of the terms. Then use the same sign as the terms.	$(+4)+(+2)=$	-8 -6 -4 -2 0 2 4 6 8
	$(-4)+(-2)=$	-8 -6 -4 -2 0 2 4 6 8

Adding Real Numbers with Unlike Signs

Verbally	Numerical Examples	Graphical Examples
To add numbers with unlike signs, first subtract the smaller absolute value from the larger absolute value. Then use the sign of the term with larger absolute value.	$(-5)+(+2)=$	-8 -6 -4 -2 0 2 4 6 8
	$(+5)+(-2)=$	-8 -6 -4 -2 0 2 4 6 8

Find each sum where each term has the same sign.

1. $10+7$ **2.** $6+14$ **3.** $14+14$

4. $-8+(-5)$ **5.** $-3+(-11)$ **6.** $-12+(-12)$

7. $2.4+5.2$ **8.** $\frac{2}{7}+\frac{2}{7}$ **9.** $\frac{1}{6}+\frac{3}{5}$

10. $-0.3+(-9.4)$ **11.** $-\frac{3}{11}+\left(-\frac{4}{11}\right)$ **12.** $-\frac{3}{10}+\left(-\frac{5}{6}\right)$

Find each sum where the terms have opposite signs.

13. $-7+12$

14. $9+(-2)$

15. $-6+6$

16. $10.8+(-4.3)$

17. $-\frac{2}{3}+\frac{5}{6}$

18. $\frac{1}{5}+\left(-\frac{1}{2}\right)$

Summary of Addition of Real Numbers

- Determine if the terms have like or unlike signs.
- If the terms have _______________ signs, add the absolute values. The sum will have the _______________ sign as both terms.
- If the terms have _______________ signs, subtract the absolute values. The sum will have the same sign as the term with the _______________ absolute value.

Find each sum.

19. $(-3)+[12+(-4)]$

20. $[-4+9]+[7+(-17)]$

Objective 2: Use the commutative and associative properties of addition.

Properties of Addition

	Algebraically	**Verbally**	**Numerical Example**
Commutative property of addition	$a+b=b+a$	The sum of two ______________ in either order is the same.	$2+5=$ $5+2=$
Associative property of addition	$(a+b)+c=a+(b+c)$	Terms can be ______________ without changing the sum.	$(1+2)+3=$ $1+(2+3)=$

21. Explain the distinction between the commutative property of addition and the associative property of addition.

22. There are two different ways to use the commutative property of addition to rewrite the following expression: $x+(y+z)$. Can you write both? ________________ and ________________

Perform the indicated operations.

23. $(5+82)+18$

24. $\frac{1}{5}+\frac{2}{5}+\left(-\frac{3}{5}\right)$

25. $0.4+(5.1+7.6)$

Objective 3: Subtract positive and negative real numbers.

It is helpful when first performing a subtraction to actually rewrite the subtraction as addition. This is not generally done once you are comfortable with performing subtraction mentally.

Subtraction

Algebraically	**Verbally**	**Numerical Example**
For any real numbers x and y, $x-y=x+(-y)$.	To subtract y from x, add the ______________ of y to x.	$6-4=6+(\_\_\_\_\_\_)$ $=\_\_\_\_\_\_$

The terms in each expression have the same sign. First rewrite each difference as a sum and then give its value.

26. $8-1$

27. $-13-(-9)$

28. $-17-(-17)$

29. $0.7-3.4$

30. $\frac{5}{9}-\frac{1}{3}$

31. $-\frac{3}{5}-\left(-\frac{3}{8}\right)$

The terms in each expression have opposite signs. First rewrite each difference as a sum and then give its value.

32. $-9-3$

33. $11-(-8)$

34. $-15-15$

35. $-7.3-5.1$

36. $-\frac{4}{9}-\frac{1}{3}$

37. $\frac{1}{5}-\left(-\frac{5}{8}\right)$

38. Determine the value of each expression.

(a) $7-2$

(b) $2-7$

(c) $-7-(-2)$

(d) $-2-(-7)$

39. We know that addition is commutative. What about subtraction?

40. The number of riders on the Top Thrill Dragster at Cedar Point Amusement park is shown in the chart below. Determine the change in the number of riders from 2003 to 2005.

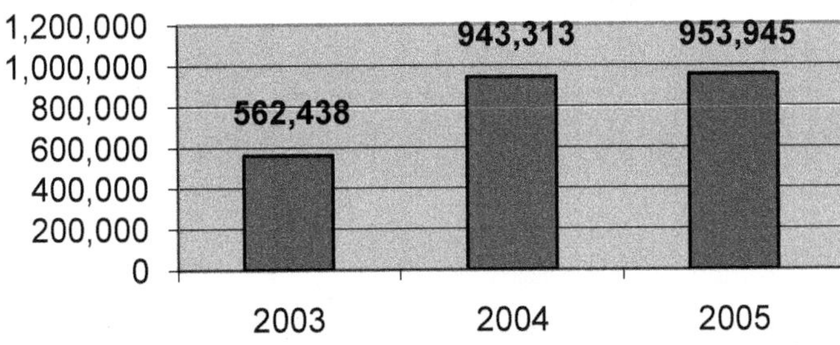

Phrases Used To Indicate Addition

Key Phrase	Verbal Example	Algebraic Example
Plus	"12 plus 8"	$12+8$
Total	"The total of \$25 and \$40"	$\$25+\$40$
Sum	"The sum of x and y"	$x+y$
Increased by	"An interest rate r is increased by 0.5%"	$r+0.005$
More than	"7 more than x"	$x+7$

Phrases Used To Indicate Subtraction

Key Phrase	Verbal Example	Algebraic Example
Minus	"x minus y"	$x-y$
Difference	"The difference between 12 and 8"	$12-8$
Decreased by	"An interest rate r is decreased by 0.5%"	$r-0.005$
Less than	"7 less than x"	$x-7$
Change	"The change from 70° to 96°"	$96^\circ-70^\circ$

Translate each verbal statement into algebraic form.

41. The total of t and six is nine.

42. Five more than z is zero.

43. The sum of a real number b and its additive inverse is zero.

44. t minus seven equals v.

45. The difference from x to y is ten.

46. w is four less than x.

47. The subtraction fact $11-7=4$ can be written as the addition fact that $4+7=$ ______.

48. The subtraction fact $25-9=16$ can be written as the addition fact ______________.

1.5 Lecture Guide: Multiplication and Division of Real Numbers

The following box contains the common notations used to indicate multiplication of the factors x and y. The times sign "×" is not included because it is not frequently used in algebra.

Notations for the Product of the Factors x and y

xy	$x \cdot y$	$(x)(y)$	$x(y)$	$(x)y$	$x * y$

Objective 1: Use the commutative and associative properties of multiplication.

Properties of Multiplication

	Algebraically	**Verbally**	**Numerical Examples**
Commutative Property	$ab = ba$	The product of two ______________ in either order is the same.	$5 \cdot 6 =$ $6 \cdot 5 =$
Associative Property	$(ab)(c) = (a)(bc)$	Factors can be ______________ without changing the product.	$(2 \cdot 3) \cdot 4 =$ $2 \cdot (3 \cdot 4) =$

1. Identify the property used to justify the equality of the two expressions in each equation below. Select from the following list:

I. Commutative Property of Addition **II.** Commutative Property of Multiplication
III. Associative Property of Addition **IV.** Associative Property of Multiplication

2. _____ $3 \cdot (4(x+2)) = (3 \cdot 4)(x+2)$

3. _____ $(x+2)(3+x) = (x+2)(x+3)$

4. _____ $3+(x+2) = 3+(2+x)$

5. _____ $(x+2)+5 = x+(2+5)$

6. _____ $(x+3)(x-5) = (x-5)(x+3)$

7. _____ $(x+5)4 = 4(x+5)$

8. Think about the difference in the meanings of the terms "regroup" and "reorder".
(a) Which term applies to the commutative property?

(b) Which term applies to the associative property?

Objective 2: Multiply positive and negative real numbers.

Multiplication of Two Real Numbers

	Algebraically	**Numerical Example**
Like signs:	Multiply the absolute values of the two factors and use a ______________ sign for the product.	$(+3)(+6)=$_____ $(-3)(-6)=$_____
Unlike signs:	Multiply the absolute values of the two factors and use a ______________ sign for the product.	$(+3)(-6)=$_____ $(-3)(+6)=$_____
Zero factor:	The product of 0 and any other factor is ______.	$(3)(0)=$_____ $(0)(-6)=$_____

9. The Signs of a Sum vs. the Signs of a Product: Fill in the correct sign of the sum or product below or indicate that not enough information is given to determine the sign.

Sum	Sign	Product	Sign
(positive) + (positive) =		(positive) · (positive) =	
(positive) + (negative) =		(positive) · (negative) =	
(negative) + (positive) =		(negative) · (positive) =	
(negative) + (negative) =		(negative) · (negative) =	
(0) + (positive) =		(0) · (positive) =	
(0) + (negative) =		(0) · (negative) =	

Calculate each product using only pencil and paper.

10. $4(-5)$

11. $-3(6)$

12. $-(-2)(-7)$

13. $3(-4)(-10)$

14. $2(-5)(-11)$

15. $-(-2)(-8)(0)$

16. $100(-13.51)$

17. $-0.1(-23.5)$

18. $\frac{1}{7}(-14)$

Product of Negative Factors

- The product is ____________________ if the number of negative factors is even.
- The product is ____________________ if the number of negative factors is odd.

The number one is called the **multiplicative identity** because 1 is the only real number with the property that $1 \cdot a = a$ and $a \cdot 1 = a$ for every real number *a*.

A Factor of 1 or −1

Algebraically	**Verbally**	**Numerical Examples**
For any real number a:		
$1 \cdot a = a$	The product of one and any real number is that same number.	$1 \cdot 5 =$ and $1 \cdot (-5) =$
$-1 \cdot a = -a$	The product of negative one and any real number is the ______________ of that real number.	$-1 \cdot 3 =$ and $-1 \cdot (-4) =$

Reciprocals or Multiplicative Inverses

Algebraically	**Verbally**	**Numerical Examples**
For any real number $a \neq 0$: $a \cdot \frac{1}{a} = 1$.	For any real number *a* other than zero, the product of the number *a* and its multiplicative inverse $\frac{1}{a}$ is 1.	$-4 \cdot \left(-\frac{1}{4}\right) = 1$, and $\frac{4}{3} \cdot \frac{3}{4} = 1$
$\frac{1}{0}$ is undefined	Zero has no multiplicative inverse.	$\frac{0}{1} = 0$ but $\frac{1}{0}$ is undefined

Give the multiplicative inverse of each of the following real numbers and then multiply the number by its multiplicative inverse.

Number	Multiplicative Inverse	Product
19. −5		
20. 4		
21. $\frac{3}{4}$		
22. $\frac{-7}{4}$		

Determine the sign of each number.

23. The multiplicative inverse of a positive number.

24. The multiplicative inverse of a negative number.

25. The additive inverse of a positive number.

26. The additive inverse of a negative number.

27. Does every real number have a multiplicative inverse?

28. Does every real number have an additive inverse?

Objective 3: Divide positive and negative real numbers.

Notations for the Quotient of x Divided by y for $y \neq 0$

$x \div y$	$\dfrac{x}{y}$	x/y	$x:y$

Relationship Between Division and Multiplication

Algebraically	Verbally	Numerical Example
For any real numbers x and y with $y \neq 0$, $x \div y = x\left(\dfrac{1}{y}\right) = \dfrac{x}{y}$	Dividing two real numbers is the same as multiplying the first number by the multiplicative ____________ of the second number.	$5 \div 2 =$

Division of Two Real Numbers

Like signs: Divide the absolute values of the two numbers and use a ______________ sign for the quotient.

Unlike signs: Divide the absolute values of the two numbers and use a ______________ sign for the quotient.

Zero dividend: $\dfrac{0}{x} = 0$ for $x \neq 0$.

Zero divisor: $\dfrac{x}{0}$ is ______________ for every real number x.

Although memorization is generally not the best way to learn mathematical concepts, it is very helpful to have the following key facts memorized when performing division.

36. The Sign of a Product vs. the Sign of a Quotient: Where possible, fill in the correct sign of each product and quotient below.

Product	Sign	Quotient	Sign
(positive) · (positive) =		(positive) ÷ (positive) =	
(positive) · (negative) =		(positive) ÷ (negative) =	
(negative) · (positive) =		(negative) ÷ (positive) =	
(negative) · (negative) =		(negative) ÷ (negative) =	
(negative or positive) · (0) =		(negative or positive) ÷ (0) =	
(0) · (negative or positive) =		(0) ÷ (negative or positive) =	

Mentally evaluate each quotient.

37. $12 \div 4$

38. $(-12) \div (-4)$

39. $-12 \div 4$

40. $12 \div (-4)$

41. $12 \div 0$

42. $0 \div 12$

43. $-24 \div \frac{1}{3}$

44. $24 \div (-3)$

45. $-234 \div 100$

46. $-234 \div (-0.001)$

47. $\frac{3}{4} \div (-12)$

48. $-\frac{5}{6} \div \left(-\frac{4}{3}\right)$

Three Signs of a Fraction:

Algebraically	**Verbally**	**Numerical Example**
For all real numbers a and b with $b \neq 0$, $-\frac{a}{b} = \frac{-a}{b} = \frac{a}{-b} = -\frac{-a}{-b}$ and $\frac{a}{b} = \frac{-a}{-b} = -\frac{-a}{b} = -\frac{a}{-b}$	Each fraction has three signs associated with it. Any two of these signs can be changed and the value of the fraction will stay the same.	$-\frac{3}{4} =$ $\frac{3}{4} =$

49. Signs of a fraction: Mentally determine the sign of each expression.

(a) $-7 \div (-8)$

Sign: ______

(b) $-\frac{-7}{8}$

Sign: ______

(c) $\frac{7}{8}$

Sign: ______

(d) $-((-7) \div (-8))$

Sign: ______

Objective 4: Express ratios in lowest terms. Any ratio can be written in fraction form. To express a ratio in lowest terms, simply reduce the fraction.

Ratio

Verbally	**Algebraically**	**Numerical Example**
The ratio of a to b is the quotient of a divided by b.	The ratio a to b can be denoted by either $a : b$ or $\frac{a}{b}$.	The ratio 5 to 8 can be denoted by either ____________ or ____________.

Write each ratio in lowest terms.

50. $12:20$

51. $60:24$

52. Twelve of 52 cards in a deck of cards are face cards. What is the ratio of face cards to all cards in the deck?

Mean and Range

The mean of a set of numerical scores is an average calculated by dividing the ____________ of scores by the number of scores, and the range of a set of scores is calculated by subtracting the ____________ score minus the ____________ score.

53. A student earned the following scores on their Beginning Algebra Exams: 77, 59, 94, 62, 71, 61. Give the range and the mean for this set of scores. Round the mean to the nearest hundredth.

Range =

Mean =

54. The price of an item was decreased from \$200 to \$170.

(a) What is the amount of the price decrease?

(b) What is the percent of the price decrease – that is, what percent of the original price is the decrease?

Phrases Used To Indicate Multiplication

Key Phrase	Verbal Example	Algebraic Example
Times	"x times y"	xy
Product	"The product of 5 and 7"	$5 \cdot 7$
Multiplied by	"The rate r is multiplied by the time t"	rt
Twenty percent of	"Twenty percent of x"	$0.20x$
Twice	"Twice y"	$2y$
Double	"Double the price P"	$2P$
Triple	"Triple the coupon value V"	$3V$

Phrases Used To Indicate Division

Key Phrase	Verbal Example	Algebraic Example
Divided by	"x divided by y"	$\frac{x}{y}$
Quotient	"The quotient of 5 and 3"	$5 \div 3$
Ratio	"The ratio of x to 2"	$x:2$

Translate each verbal statement into algebraic form.

55. a times six is equal to twelve.

56. The product of p and q is equal to the product of q and p.

57. Twice x is greater than six.

58. The ratio of three to x is equal to ten.

59. z divided by two is equal to three less than z.

60. The quotient of seven and nine is equal to the multiplicative inverse of x.

1.6 Lecture Guide: Natural Number Exponents and Order of Operations

Objective 1: Use natural number exponents.

Repeated Multiplication: Exponential notation is a concise notation to indicate repeatedly multiplying the same factor a given number of times. The expression $2\cdot2\cdot2\cdot2\cdot2\cdot2\cdot2\cdot2$ can be written using exponential notation. The base, or factor being repeatedly multiplied, is 2. The exponent, or number of times the factor is repeated, is 8. So $2\cdot2\cdot2\cdot2\cdot2\cdot2\cdot2\cdot2=2^8$. Exponential notation is a concise notation to indicate repeatedly multiplying the same factor a given number of times.

Exponential Notation

Algebraically	**Verbally**	**Numerical Examples**
For any natural number n, $b^n = \underbrace{b\cdot b\cdot\ldots\cdot b}_{n \text{ factors of } b}$ with base b and exponent n.	For any natural number n, b^n is the product of b used as a factor n times. The expression b^n is read as "b to the nth power."	$5^3 =$ $(-4)^2 =$

Write each expression in exponential form.

1. $4\cdot4\cdot4\cdot4\cdot4$

2. $x\cdot x\cdot x$

Write each expression in expanded form.

3. $(-2)^4$

4. $(5x)^3$

To avoid errors it is very important to identify the base of an exponential expression. If an exponent is on a number or variable, then that number or variable is the base. If an exponent is outside a pair of grouping symbols, then the contents of this pair of grouping symbols is the base. Identify the correct base, exponent and value of each expression.

Exponential Expression	**Base**	**Exponent**	**Expanded Form**
5. -4^2			
6. $(-4)^2$			
7. $3x^4$			
8. $(3x)^4$			

Mentally evaluate each expression.

9. 6^2

10. 2^6

11. $\left(\frac{2}{3}\right)^4$

12. -5^2

13. $(-5)^2$

14. $(-1)^{21}$

15. Can you explain the subtle distinction between $(-3)^4$ and -3^4 ?

16. $(-5)^3$ and -5^3 have the same value although the bases are different. Can you explain this?

Objective 2: Use the standard order of operations.

17. (a) Which of the following do you think is the correct evaluation of the expression $6+3\div3-7\cdot4$?

Option I

$$\begin{aligned} 6+3\div3-7\cdot4 &= 9\div3-7\cdot4 \\ &= 6-7\cdot4 \\ &= -1\cdot4 \\ &= -4 \end{aligned}$$

Option II

$$\begin{aligned} 6+3\div3-7\cdot4 &= 6+1-28 \\ &= 7-28 \\ &= -21 \end{aligned}$$

Option III

$$\begin{aligned} 6+3\div3-7\cdot4 &= 9\div(-4)\cdot4 \\ &= \frac{9}{-4}\cdot4 \\ &= -9 \end{aligned}$$

(b) Try evaluating the above expression on your calculator. Do you agree with the calculator result?

A major objective in this section is to master the order of operations. The order of operations gives a consistent method for evaluating mathematical expressions like the one above.

Standard Order of Operations

Step 1: Start with the expression within the innermost pair of __________________ __________________.

Step 2: Perform all __________________.

Step 3: Perform all __________________ and __________________ as they appear from left to right.

Step 4: Perform all __________________ and __________________ as they appear from left to right.

Some common grouping symbols are [ ], { }, ( ), | |, $\sqrt{\ }$, and —. The first four grouping symbols contain both a beginning symbol before and an ending symbol after the group. In the radical symbol $\sqrt{\ }$ and the fraction bar — the group is determined by the length of the horizontal bar.

Use the standard order of operations to perform the indicated operations. Use a calculator only to check your results.

18. $-(-3)^2$

19. 0^8

20. -1^8

21. $(-1)^8$

22. $12+4\cdot 3$

23. $(12+4)\cdot 3$

24. $12\div 4(3)$

25. $12\div(4\cdot 3)$

26. $(7+5)^2$

27. 7^2+5^2

28. $28-20\div 4+6$

29. $(28-20)\div 4+6$

Use the standard order of operations to perform the indicated operations. Use a calculator only to check your results.

30. $\dfrac{4(3) \div 3(-4)}{2}$

31. $36 \div 9 \cdot 3 \div 2$

32. $7 - 7(4^2 - 13)$

33. $\dfrac{5^2 - 4^2}{\sqrt{5^2 - 4^2}}$

34. $5 - [8 - (6 - 9)]$

35. $5 - 3[4 - 6(5 - 1)]$

36. $\dfrac{3^2 - 4^2}{20 - 2(9 - 6)}$

37. $\dfrac{4 + \sqrt{25 - 4(1)(4)}}{2}$

Objective 3: Use the distributive property of multiplication over addition. The distributive property can be used in two ways: to expand expressions that are in a factored form and to factor expressions with terms that have a common factor.

Distributive Property of Multiplication Over Addition

Algebraically	**Verbally**	**Numerical Examples**
For all real numbers a, b, and c, $a(b+c)=ab+ac$ and $(b+c)a=ba+ca$.	Multiplication distributes over addition.	$3(4+5)=3\cdot4+3\cdot5$ $(4+5)3=4\cdot3+5\cdot3$

Use the distributive property to expand each expression in the first column and to factor each expression in the second column.

Expand

38. $6(3x+5)$

40. $(4x+7y)5$

42. $-1(2-3x)$

Factor

39. $-2x-10$

41. $18x+24y$

43. $-x+2$

44. What property justifies the fact that $a(b+c)=(b+c)a$?

45. What property justifies the fact that $a(b+c)=a(c+b)$?

46. What property justifies the fact that $a(b+c)=ab+ac$?

Objective 4: Simplify expressions by combining like terms.

Adding Like Terms: Use the distributive property to combine like terms.

47. $7x+4x$

48. $8x-6x$

49. $3x+7y+11x$

50. $x+3y-13x$

51. $(2x+7y)+(3x-4y)$

52. $(3x-5y)-(2x-7y)$

53. $4(x-3)+5(3x-7)$

54. $8(x-3)-2(2x+1)$

Phrases Used To Indicate Exponentiation

Key Phrases	Verbal Examples	Algebraic Examples
To a power	"3 to the 6th power"	3^6
Raised to	"y raised to the 5th power"	y^5
Squared	"4 squared"	4^2
Cubed	"x cubed"	x^3

Translate each verbal statement into algebraic form.

55. x raised to the fourth power

56. The square of the quantity x plus two.

Write each expression in the horizontal one-line format used by calculators.

57. $\dfrac{12+8}{4-9}$

58. $\dfrac{10-5^2}{3}$

Each expression is given in the horizontal one-line format used by calculators. Rewrite each expression in the standard algebraic format.

59. $x \wedge 2-4/x-2$

60. $(x \wedge 2-4)/(x-2)$

1.7 Lecture Guide: Using Variables and Formulas

Objective 1: Evaluate an algebraic expression for specific values of the variables.

Evaluate $x^2 - 4x - 5$ for each value of x. (Hint: See Technology Perspective 1.7.1 for using a calculator or a spreadsheet to check your work.)

1. $x = 3$

2. $x = 5$

3. $x = -2$

Evaluate the following expressions for $x = 4$ and $y = 9$.

4. $-|y - x|$

5. $\sqrt{y} + 1 - \sqrt{x}$

6. $5x + 3y - 10$

Evaluate the following expressions for $x = -3$ and $y = -5$.

7. $\dfrac{x + 4y}{x - 4y}$

8. $\dfrac{x^2 - 4y^2}{x^2 + xy - 6y^2}$

Objective 2: Use algebraic formulas. Algebraic formulas are used in nearly all areas of mathematics, business, and the sciences. A formula describes a relationship between specific variables. For example, the area A of a triangle is given by the formula $A = \frac{1}{2}b \cdot h$, where b represents the length of the base of the triangle and h represents the height of the triangle. This relationship holds true for all triangles.

Find the area of each triangle. Remember, area is given in square units.

9. $A =$

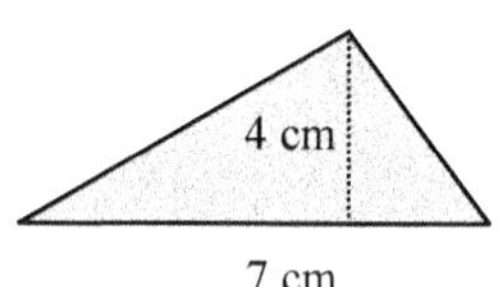

10. $A =$

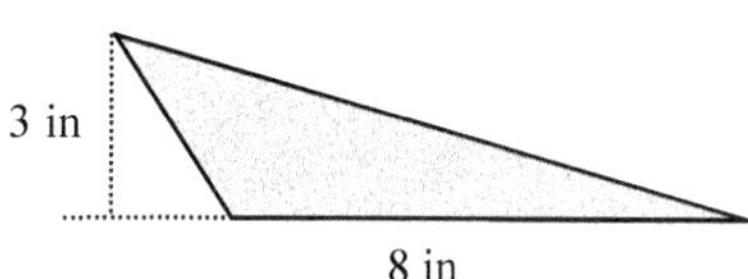

11. The formula for Fahrenheit temperature is given by $F = \frac{9}{5}C + 32$. Find the Fahrenheit temperature if the Celsius temperature C is $55°$.

12. The formula for the amount in a bank account paying a simple interest rate R for T years is given by $A = P + PRT$, where P is the principal or initial amount. Find the amount in a bank account after 1 year if there was an initial deposit of \$5,000 and the account earned 5% simple interest.

13. The formula for the perimeter of a rectangle is given by $P = 2l + 2w$. Find the perimeter P of a rectangle if the length l is 35 meters and the width w is 20 meters.

Objective 3: Use subscript notation.

One common usage of subscript notation is the slope formula, $m = \frac{y_2 - y_1}{x_2 - x_1}$, which will be developed in Chapter 3. This formula is used to calculate *m*, the slope of a line that passes through the points (x_1, y_1) and (x_2, y_2).

14. Use this formula to calculate the value of *m* for the line through the points $(x_1, y_1) = (3,5)$ and $(x_2, y_2) = (7,8)$.

15. Use this formula to calculate the value of *m* for the line through the points $(x_1, y_1) = (-1,4)$ and $(x_2, y_2) = (2,-3)$.

A **sequence** is an ordered set of numbers with a first number, a second number, a third number, etc. Subscript notation often is used to denote the terms of a sequence: a_1, a_2, and a_n. These terms are read *a sub one, a sub two,* and *a sub n*, respectively. If a sequence follows a predictable pattern, then we may be able to describe this pattern with a formula for a_n. Consider the sequence 5, 4, 3, 2. Here, $a_1 = 5$, $a_2 = 4$, $a_3 = 3$, and $a_4 = 2$.

Use each formula to calculate the first three terms (a_1, a_2, and a_3) of each sequence.

16. $a_n = 2n - 5$

17. $a_n = 5n - 8$

Objective 4: Check a possible solution of an equation. A solution of an equation is a value for the variable that satisfies the equation. This means that when the value is substituted for the variable, the expressions on each side of the equation will have the ____________ value.

Check whether each indicated value of x is a solution of the given equation.

18. $x - 1 = 2x - 4$

(a) Check $x = 3$

(b) Check $x = 5$

19. $x^2 + 3x - 10 = 0$

(a) Check $x = 3$

(b) Check $x = 5$

20. $\dfrac{3x-2}{x+4} = \dfrac{2}{3}$

(a) Check $x = -4$

(b) Check $x = 2$

21. $(4x+3)(x-2) = 0$

(a) Check $x = -\dfrac{4}{3}$

(b) Check $x = -\dfrac{3}{4}$

1.8 Lecture Guide: Geometry Review

Objective 1: Classify and Name polygons.

A **polygon** is a closed geometric figure formed by straight line segments. Some common polygons are given below.

Triangle: ____________ sides	**Quadrilateral:** ____________ sides	**Trapezoid:** quadrilateral with two ____________ sides parallel	**Parallelogram:** quadrilateral with opposite sides ____________	**Rhombus:** parallelogram with four ____________ sides
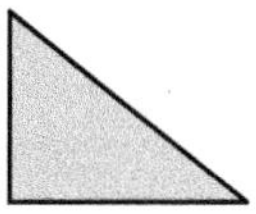	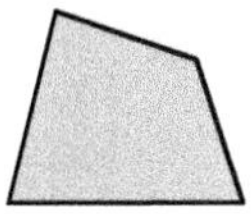	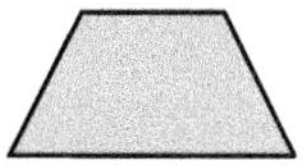		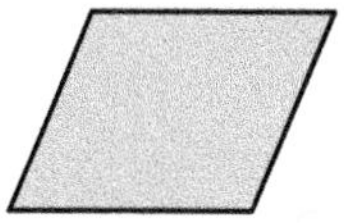

Rectangle: parallelogram with adjacent sides ____________	**Square:** rectangle with four ____________ sides	**Pentagon:** ____________ sides	**Hexagon:** ____________ sides	**Octagon:** ____________ sides
		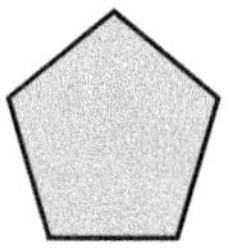	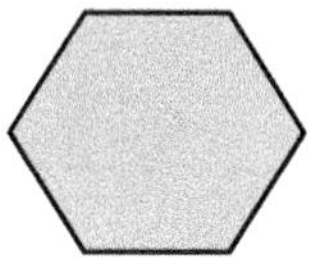	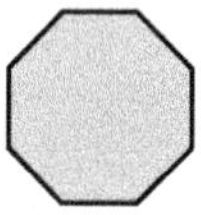

As a reminder, note the difference between units of length, area, and volume shown below.

Sample measurements:

Length: ____ 1 cm

Area: 1 cm^2

Volume: 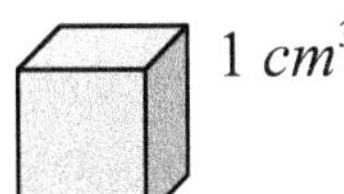 1 cm^3

Objective 2: Calculate the perimeter of common geometric figures.

1. **Perimeter of a Square**
   Each side of the square shown is 3.2 inches long. Determine its perimeter.

3.2 in

2. **Perimeter of a Rectangle**
   Determine the perimeter of this rectangle.

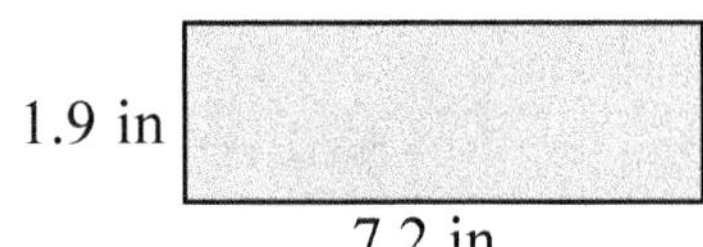

3. **Perimeter of a Triangle**
   Determine the perimeter of this triangle.

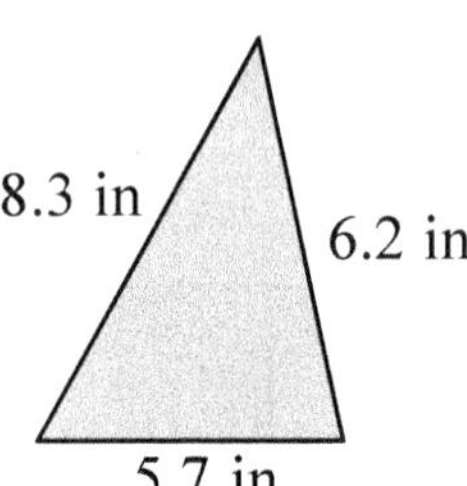

4. **Circumference of a Circle**
   (a) Determine the exact circumference of this circle.

   (b) Approximate this length to the nearest tenth of an inch.

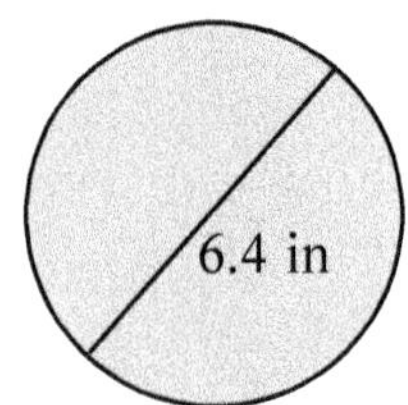

Objective 3: Calculate the area of common geometric regions.

5. Area of a Rectangle

Determine the area of this rectangle.

8 in

20 in

6. Area of a Region

If $\frac{1}{3}$ of this circular region is to be painted, determine to the nearest tenth of a square inch the area to be painted.

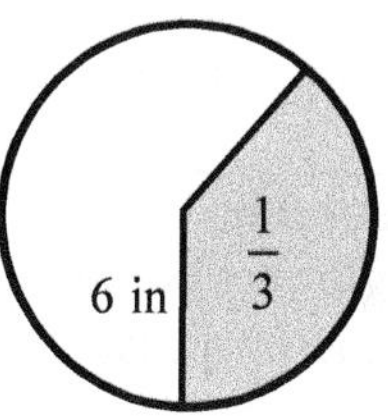

Objective 4: Calculate the volume of a rectangular solid.

7. Volume of a Crate

Determine the volume of a rectangular crate used to ship car parts. The crate measures 18 inches wide, 24 inches long, and 20 inches deep.

8. Volume of an Exercise Ball

Determine the volume of an exercise ball with a diameter of 22 inches. Use the formula $V = \frac{4}{3}\pi r^3$ to approximate the volume to the nearest tenth of a cubic inch.

Objective 5: Classify and name common angles.

Acute Angle: between 0° and 90°	**Right Angle:** 90°	**Obtuse Angle:** between 90° and 180°	**Straight Angle:** 180°

Complementary and Supplementary Angles

Verbally	**Algebraically**	**Graphically**
Complementary: Two angles are complementary if the sum of their measures is ______.	$m\angle a + m\angle b = 90^\circ$	b a
Supplementary: Two angles are supplementary if the sum of their measures is ______.	$m\angle a + m\angle b = 180^\circ$	b a

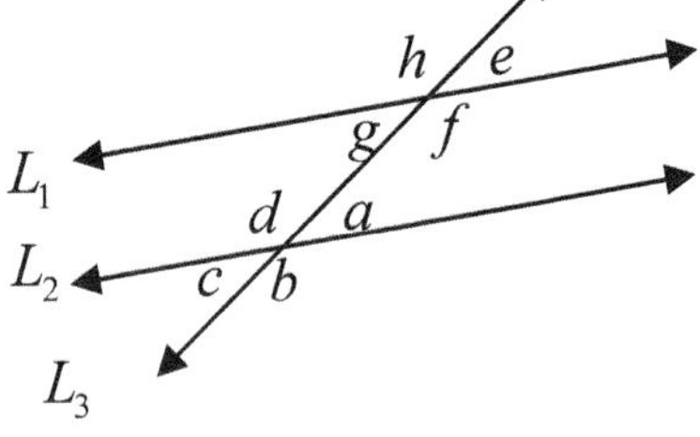

In problems 9-18 match each description with the most appropriate choice. (Some choices may be used multiple times while others may not be used.) Assume that L_1 is parallel to L_2 in the figure to right.

9. An angle that measures between 0° and 90°. ____

10. An angle that measures between 90° and 180°. ____

11. An angle that measures 180°. ____

12. The measure of a right angle. ____

13. The sum of the measures of two supplementary angles. ____

14. The sum of the measures of two complementary angles. ____

15. Angles *e* and *f*. ____

16. Angles *e* and *g*. ____

17. Angles *b* and *h*. ____

18. Angles *a* and *g*. ____

A. Acute angle

B. Adjacent angles

C. Alternate exterior angles

D. Alternate interior angles

E. Obtuse angle

F. Straight angle

G. Vertical angles

H. 0°

I. 90°

J. 180°

In problems 19-25 assume that $m\angle a = 31^\circ$ and that L_1 is parallel to L_2 in the figure.

Determine the measure of each of the following angles.

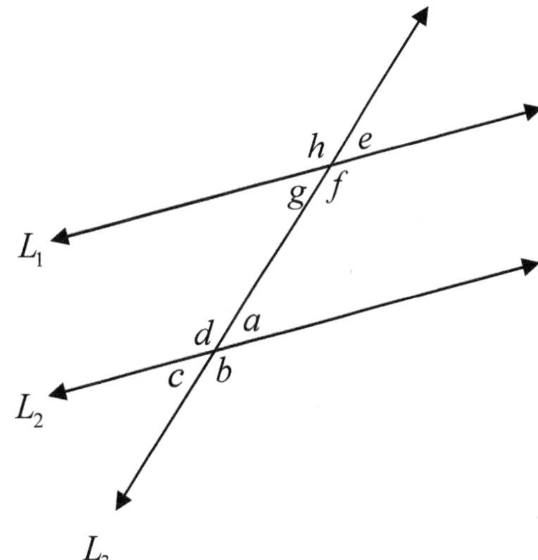

19. $\angle b$

20. $\angle c$

21. $\angle d$

22. $\angle e$

23. $\angle f$

24. $\angle g$

25. $\angle h$

Lecture Guides for

Chapter 2

Linear Equations and Patterns

2.1 Lecture Guide: The Rectangular Coordinate System

Objective 1: Plot ordered pairs on a rectangular coordinate system.

1. Identify and label:
(a) x**-axis**
(b) y**-axis**
(c) origin
(d) quadrants

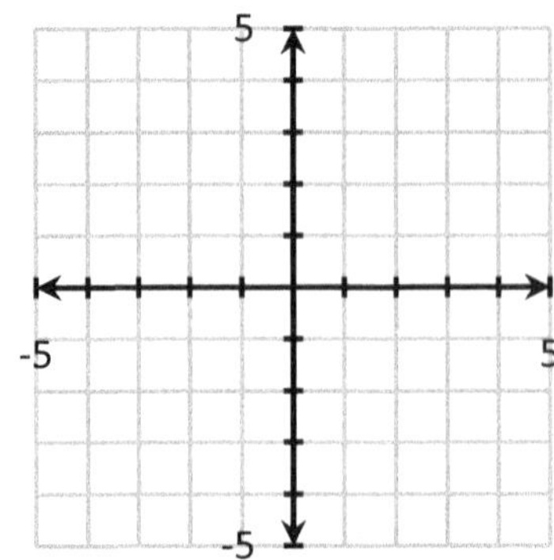

2. Note that an ordered pair consists of two ____________________, an x-____________________ and a y-____________________. The one that is always listed first is the ____________________.

3. Use the graph in question 1 to fill in each blank.
(a) Moving up or down on the coordinate plane changes the ___-coordinate but not the ___-coordinate.
(b) Moving left or right on the coordinate plane changes the ___-coordinate but not the ___-coordinate.
(c) Every point on the x-axis has a y-coordinate of ______ and every point on the y-axis has an x-coordinate of ______.

4. Identify the coordinates of points A, B, C, and D. Give the quadrant in which each point is located.

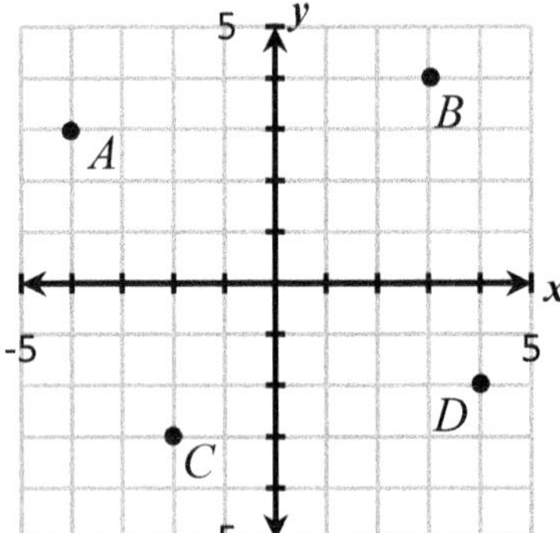

(a) A:

(b) B:

(c) C:

(d) D:

5. Identify the coordinates of one point that would lie between Quadrants II and I.

6. Identify the coordinates of one point that would lie between Quadrants IV and I.

7. Plot and label the points whose coordinates are given on the following coordinate system.

(a) $A(-2, 4)$

(b) $B(5, -3)$

(c) $C(-3, -4)$

(d) $D(-2, 0)$

(e) $E(0, 3)$

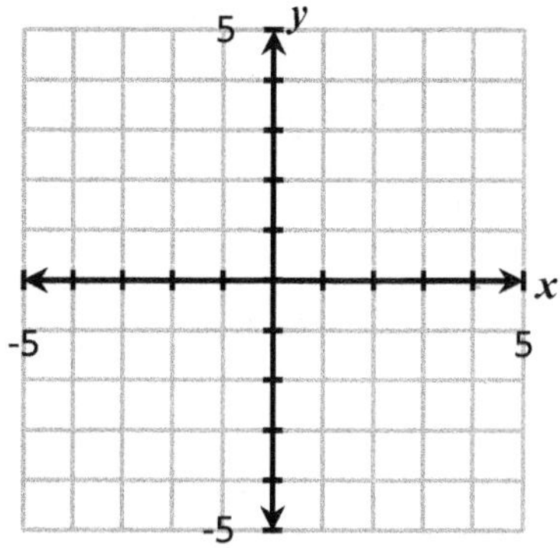

Objective 2: Draw a scatter diagram of a set of points.

A **scatter diagram** for a set of data points is simply a graph of these points.

8. Draw a scatter diagram for the following set of points:

x	-6	-3	0	3	6
y	6	4	2	0	-2

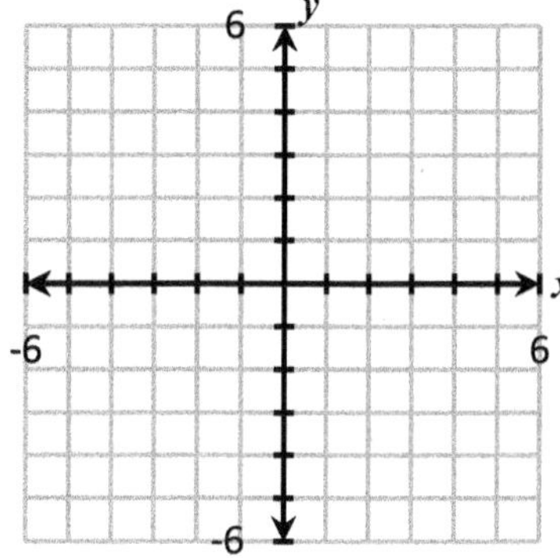

Do these points lie on a straight line?

Objective 3: Identify data whose graph forms a linear pattern.

Arithmetic Sequences

Numerically: An arithmetic sequence has a ______________ change, d, from term to term.

Graphically: The distinct points of the graph of an arithmetic sequence form a ______________ pattern. There is a ______________ change in height between consecutive points.

9. Rewrite the sequence $-5, -1, 3, 7, 11$ using subscript notation.

$a_1 =$ ______, $a_2 =$ ______, $a_3 =$ ______, $a_4 =$ ______, $a_5 =$ ______

10. When creating a graph of a sequence, use n as your ______-value and use ______ as your y-value.

11. Represent the following sequence using ordered subscript notation and ordered pair notation. Then Complete the table and the graph.

Sequence: $3,\ -1,\ -5,\ -9,\ -13$

Table

n	a_n

Graph

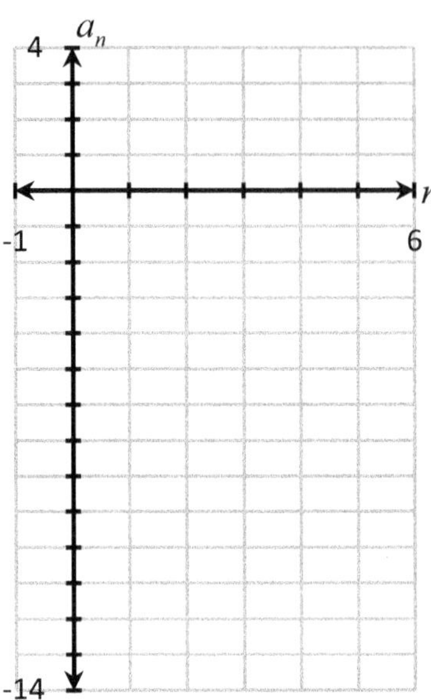

Subscript notation:

Ordered pair notation:

12. Consider the given graph of the sequence.

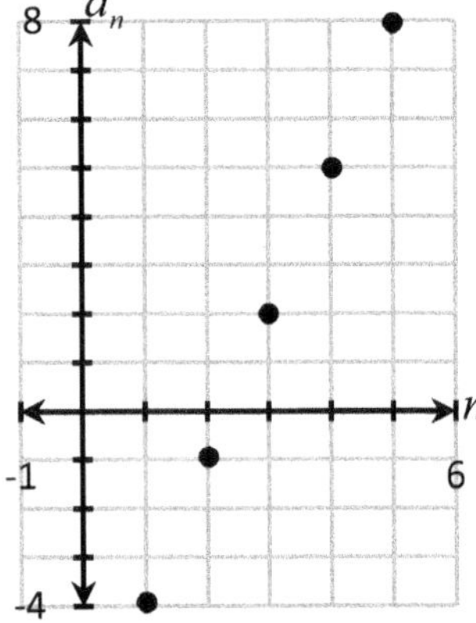

(a) Write the first five terms of this sequence.

(b) Is this sequence arithmetic?

(c) If this sequence is arithmetic, determine the common difference d of this sequence.

13. Consider the given graph of the sequence.

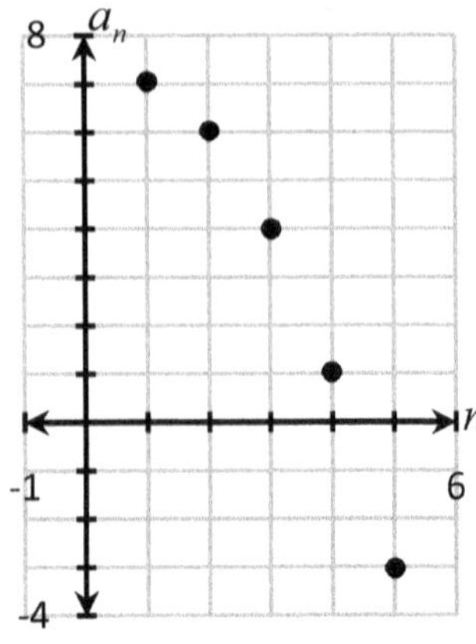

(a) Write the first five terms of this sequence.

(b) Is this sequence arithmetic?

(c) If this sequence is arithmetic, determine the common difference d of this sequence.

14. It is important to observe that the graph of an ***arithmetic*** sequence forms a ________________ pattern.

Determine whether each sequence is an arithmetic sequence and whether its graph forms a linear pattern.

15. 4, 8, 12, 16, 20

16. 4, 8, 16, 32, 64

17. The equation $a_n = 300n + 1000$ gives the total payments in dollars after n months on a loan for a new truck. Calculate and interpret each value of a_n.

(a) a_0

(b) a_{12}

(c) a_{24}

Objective 4: Interpret a line graph.

18. The graph shown below gives the altitude of a small airplane at a given time. The time x is given in minutes from the start of the flight and the altitude y is given in thousands of feet. Answer each question by examining this graph.

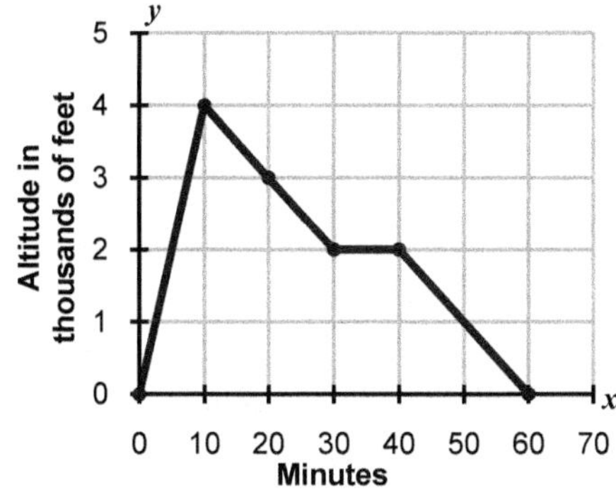

(a) What was the highest altitude reached by the plane?

(b) How long after the flight began did the plane reach this highest altitude?

(c) During what time was the plane flying level?

(d) How long was the flight?

2.2 Lecture Guide: Function Notation and Linear Functions

Objective 1: Use function notation.

Function Notation

The notation $f(x)$ is referred to as function notation and is read "______ of ______" or "$f(x)$ is the ____________ value for an ____________ value of x."

1. Given $f(x) = -4x + 3$, evaluate each of the following:

(a) $f(7)$

(b) $f(0)$

(c) $f(-2)$

(d) $f\left(-\frac{3}{2}\right)$

Objective 2: Use a linear equation to form a table of values and to graph a linear equation.

2. Use the function $f(x) = 3x - 4$ to complete the following table and graph the line.

Table

x	$f(x)$
−2	
−1	
0	
1	
2	

Graph

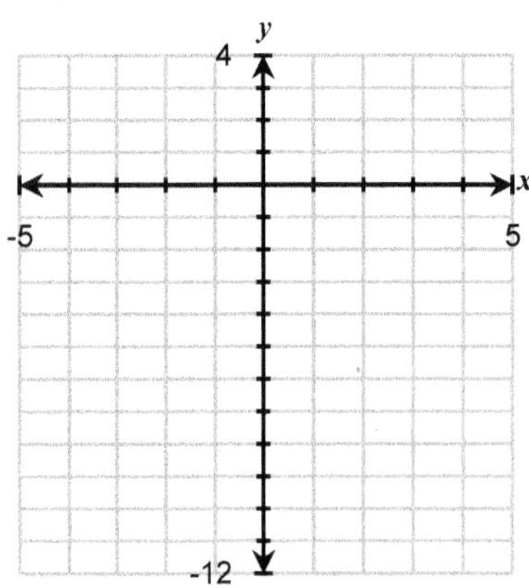

3. The function $f(x) = 3x - 4$ from problem 2 is called a linear function because its graph is a ________________ ________________. Functions in the form $f(x) = mx + b$ are called linear functions.

4. Use the function $f(x) = x - 1$ and a calculator or a spreadsheet to complete the table below.

x	$f(x)$
−3	
−2	
−1	
0	
1	
2	
3	

5. Use the function $f(x) = -2x + 5$ and a calculator or a spreadsheet to complete the table below.

x	$f(x)$
0	
10	
20	
30	
40	
50	
60	

Objective 3: Write a linear function to model an application.

6. Consider a car loan with payments of \$200 per month and a down payment of \$700.
(a) Give a function that models the total paid by the end of the xth month.
$f(x) =$ ________________
(b) Give the total paid by the end of the 36th month.

7. If you make two investments totaling \$3,000 and x represents the amount of one investment, write a function that represents the amount in the other investment. Then complete the table of values.

x Amount of first investment	$f(x) =$ ________ Amount of second investment
1,000	
1,500	
2,300	

8. If you have a 10-foot board that is to be cut in two pieces, and x represents the length of one of the pieces, write a function that represents the length of the other piece. Then complete the table of values.

x Length of first piece	$f(x) =$ ________ Length of second piece
1	
7	
8	

9. You have 40 feet of fencing to enclose three sides of a rectangular pen, and x represents the amount of fencing used for the width of the pen. Write a function that represents the amount of fencing remaining for the length of the pen. Then complete the table of values.

x Width	$f(x) =$ ________ Length
5	
10	
15	

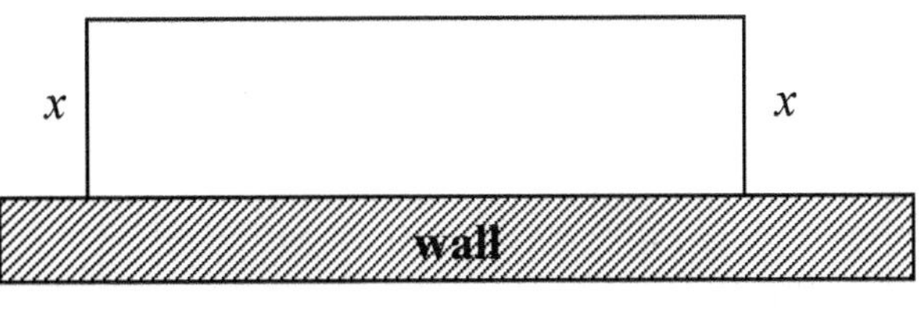

10. An airplane has a speed of x mi/h in calm skies.

(a) Write a function in terms of x for the rate of this airplane traveling in the same direction as a 20-mi/h wind.

(b) Write a function in terms of x for the rate of this airplane traveling in the opposite direction of a 20-mi/h wind.

(c) Write a function in terms of x for the distance the airplane travels in 3 hours going in the same direction as a 20-mi/h wind.

11. If you have 2 gallons of insecticide to which you are planning to add some water to dilute the mixture. Letting x represent the number of gallons of water that you add, write a function that represents the total volume in gallons of the mixture. Then complete the table of values.

x Gallons of water	$f(x) =$ ________ Total volume
5	
8	
12	

12. The price of every item in a store has been marked down by 10%. Let x represent the original price of an item.

(a) Write a function for the new price of an item with an original price of x dollars.

(b) Complete the following table for the new price of each item whose original price is given.

x Original price	$f(x) =$ ________ New price
22	
45	
80	

13. Use the given graph to complete the table.

Graph

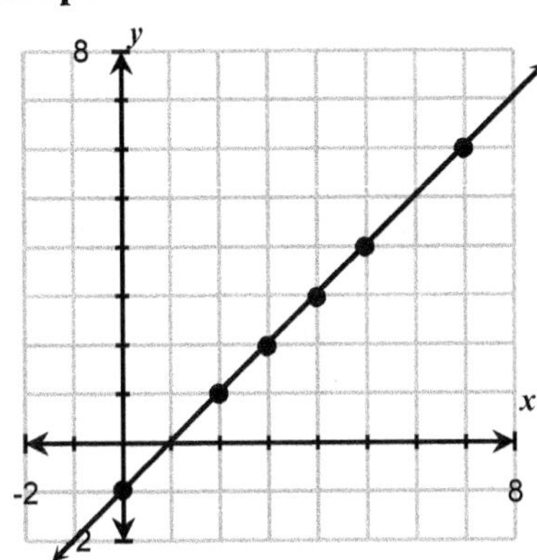

Table

x	$f(x)$
0	
2	
3	
	3
	4
	6

14. Use the given table of values to determine the missing input and output values.

Table

x	$f(x)$
−1	−2
0	0
1	2
2	4
3	6
4	8

(a) $f(2) =$ ______
(If the input value is 2, what is the output value?)

(b) $f(4) =$ ______

(c) $f(x) = 2$ for $x =$ ______
(What is the input value if the output value is 2?)

(d) $f(x) = 4$ for $x =$ ______

15. Use the given graph to determine the missing input and output values.

Graph

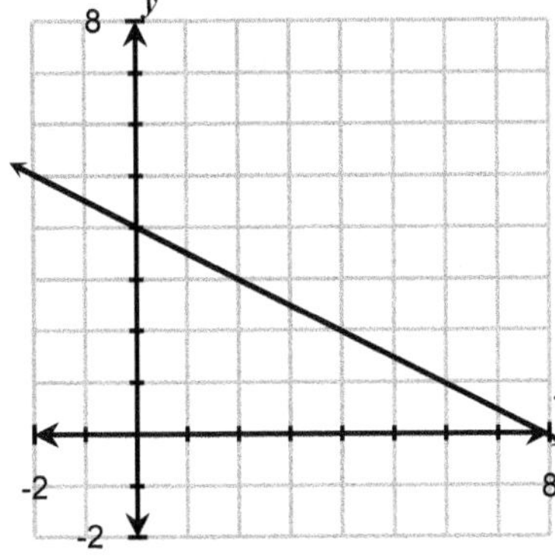

(a) $f(0) =$ ______
(If the input value is 0, what is the output value?)

(b) $f(2) =$ ______

(c) $f(x) = 0$ for $x =$ ______
(What is the input value if the output value is 0?)

(d) $f(x) = 2$ for $x =$ ______

2.3 Lecture Guide: Graphs of Linear Equations in Two Variables

Objective 1: Check possible solutions of a linear equation.

1. Before we check the solution of a linear equation in two variables, we will review how we check a solution of a linear equation in one variable.

(a) Is $x=6$ a solution of $3x-9=x+5$?

(b) Is $x=7$ a solution of $3x-9=x+5$?

We will now check a solution of a linear equation in two variables.

Solution of a linear equation $y=mx+b$

A solution of a linear equation of the form $y=mx+b$ is an ordered pair (x,y) that makes the equation a ____________ statement.

2. Test each ordered pair to determine whether it is a solution of the equation $y=2x-1$.

(a) $(-1,-3)$

(b) $(2,1)$

(c) $(0,-1)$

3. Any ordered pair that is a ________________ of a linear equation will lie on the ________________.

4. Consider the graph below.

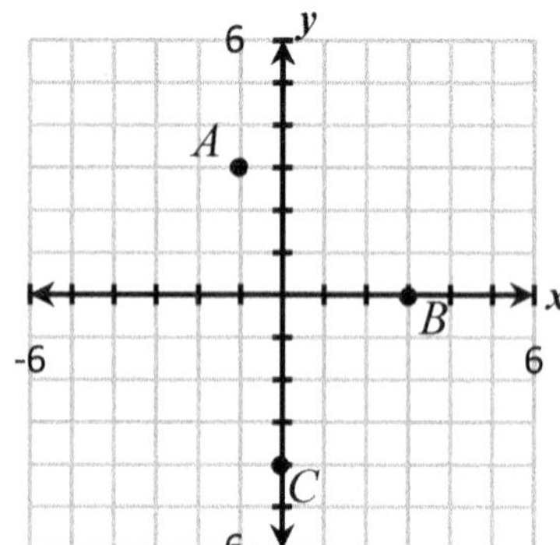

Determine whether the points *A*, *B*, and *C* on the graph are solutions of $y=\frac{4}{3}x-4$.

Objective 2: Determine the intercepts from a graph.

x- and y- intercepts

The x-intercept of a graph is the point with a y-coordinate of ______.
The y-intercept of a graph is the point with an x-coordinate of ______.

5. Determine the intercepts of the graph below.

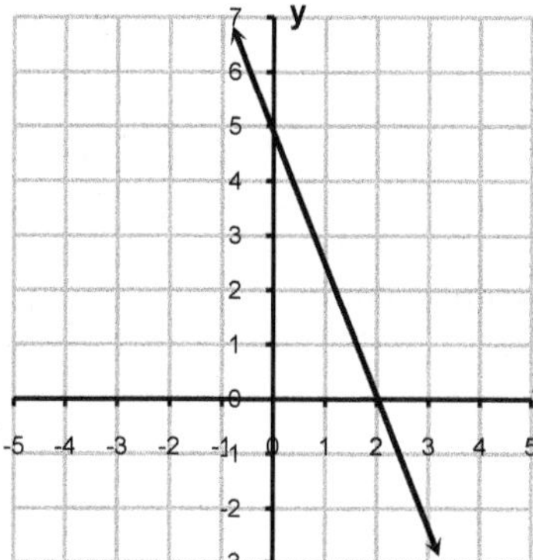

(a) x-intercept: ___________

(b) y-intercept: ___________

6. In the graph below the input x is the number of units produced by a machine in a factory. The output y is the profit made by the sale of these units when they are produced. Determine the intercepts and interpret the meaning of each.

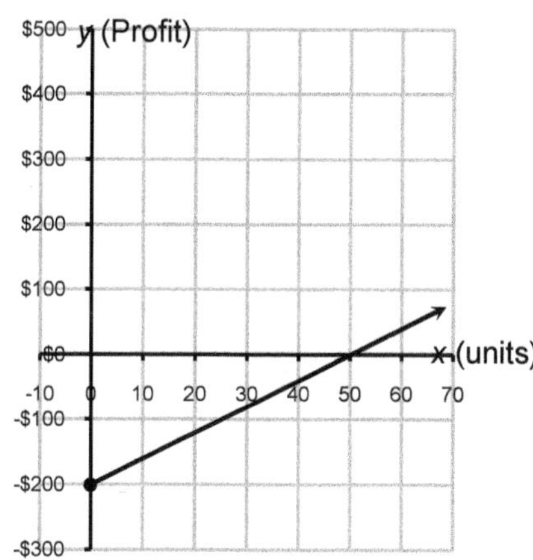

(a) x-intercept: ___________

(b) Meaning of x-intercept:

(c) y-intercept: ___________

(d) Meaning of y-intercept:

7. Use the table for the linear equation $y = mx + b$ to identify the x- and y-intercepts of the graph of the line through these points.

x	$f(x)$
-1	-6
0	-3
1	0
2	3
3	6
4	9

(a) x-intercept: ___________

(b) y-intercept: ___________

8. If a car gets 30 miles per gallon and travels 60 miles per hour, then the car uses 2 gallons of gas every hour. Assuming the car's gas tank starts with 15 gallons, write a function that represents the number of gallons of gas remaining after x hours is $f(x) =$ ________ $x +$ _______ .

(a) Use your calculator to graph this equation using a window of $[0,10,1]$ by $[0,15,5]$. Draw a rough sketch of your calculator graph below.

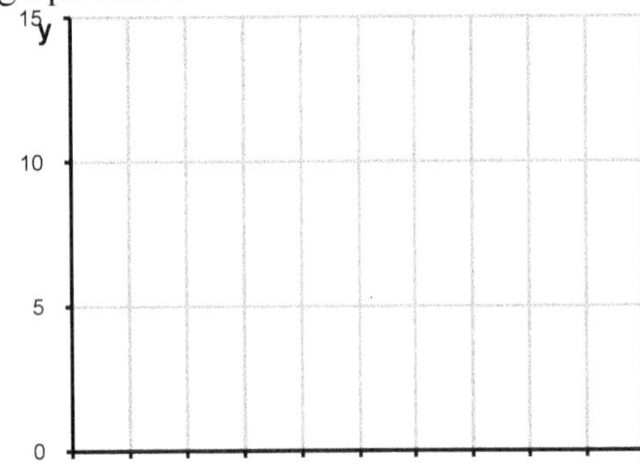

(b) Press **TRACE**. Type **0** and **ENTER**. The y-intercept is ________________.

(c) Meaning of y-intercept:

(d) Then type **7.5** and **ENTER**. The x-intercept is ________________.

(e) Meaning of x-intercept:

Objective 3: Determine the point where two lines intersect.

When we refer to two or more equations at the same time we refer to this as a system of equations. A point where two lines intersect is called a ____________ of a system of linear equations. This point of intersection is an ordered pair that makes both equations ____________ at the same time.

9. **(a)** Check each point to test if it is a solution of each equation:

Point	**Solution of $y = -x + 7$? (yes/no)**	**Solution of $y = 3x - 5$? (yes/no)**
$(6,1)$		
$(1,-2)$		
$(3,4)$		

(b) Which point would you conclude is a solution of both equations?

(c) Enter $Y_1 = -x + 7$ and $Y_2 = 3x - 5$ using a window of $[-10,10,1]$ by $[-10,10,1]$ on your calculator and press **2nd, TRACE, 5, ENTER, ENTER, ENTER.** Does this support your conclusion?

10. Consider the graph of the system: 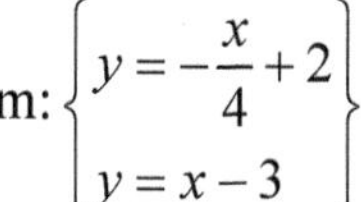

$$\begin{cases} y = -\dfrac{x}{4} + 2 \\ y = x - 3 \end{cases}$$

(a) The solution of the system of equations shown is .

(b) Verify that this point satisfies both equations.

(c) Use the **Intersect** feature on your calculator to check this solution. (See Technology Perspective 2.3.2 for help.)

A common expression is "Two points determine a ________________ ________________." To obtain the graph of a linear equation, we can find any two points that satisfy the equation and then draw a line through these points. It is usually a good idea to find a third point as a check.

11. Consider the equation $y = \dfrac{x}{2} + 1$. Plot the y-intercept and one other point to graph this line. Select a third point to double-check your work.

Table:

x	y

Graph:

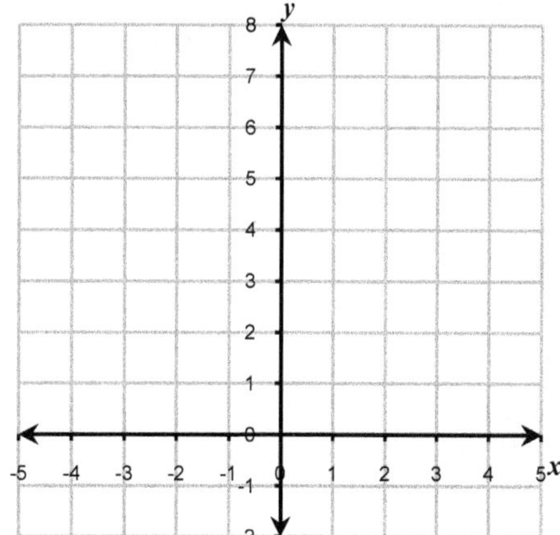

12. Consider the equation $y = -2x + 4$. Plot the y-intercept and one other point to graph this line. Select a third point to double-check your work.

Table:

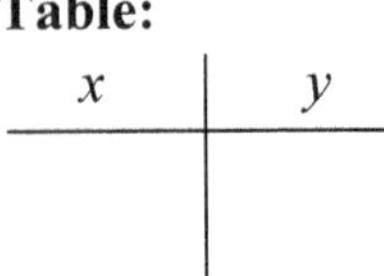

Graph:

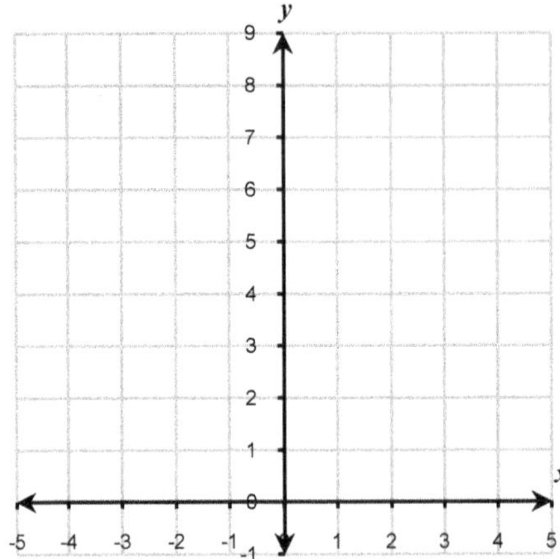

13. Solve the system of linear equations $\begin{cases} y = 3x - 5 \\ y = -x + 7 \end{cases}$ by graphing each equation on the same coordinate system and determining the point of intersection. Check the coordinates of this point in both of the linear equations.

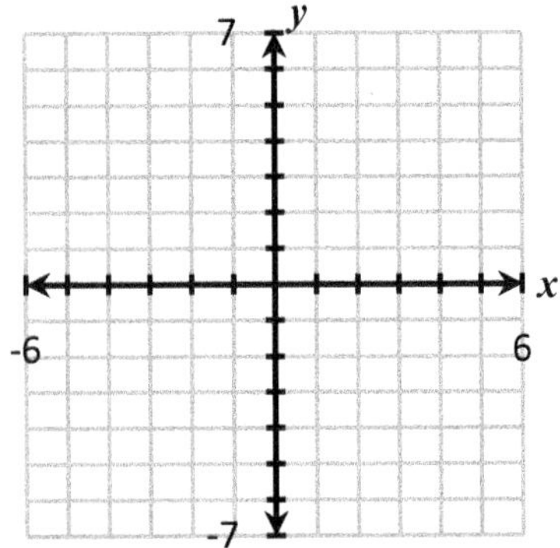

14. Use a calculator to complete the table of values for $y = 3x - 6$ and then identify the x- and y-intercepts of the graph of the line through these points.

Table

x	$f(x)$
-3	
-2	
-1	
0	
1	
2	
3	

(a) x-intercept: ____________

(b) y-intercept: ____________

15. Use a calculator with a **Trace** feature to graph the linear function $y = 1.5x - 9$ on the window [10, 10, 1] by [10, 10, 1], and then identify the *x*- and *y*-intercepts of this line. (*Hint:* See Technology Perspective 2.3.1.)

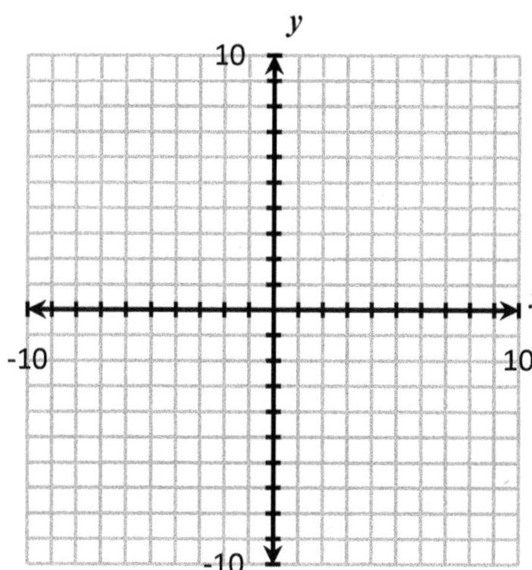

(a) *x*-intercept: ___________

(b) *y*-intercept: ___________

16. Suppose you plan to rent a car and have two rental companies to choose from. The first company offers a rate of $40 per day plus 10 cents per mile, and the second company offers a rate of $25 per day plus 25 cents per mile. The cost for a two-day rental would be determined by the following equations:

Company A: $y_1 = 0.10x +$ ______

Company B: $y_2 =$ ______ $x + 50$

(a) Use your calculator to graph each equation using a window of $[0, 500, 100]$ by $[0, 200, 50]$. Draw a rough sketch of your calculator graph below.

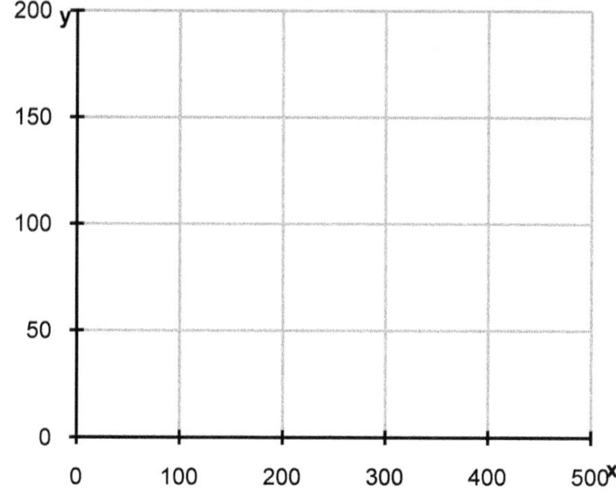

(b) Use your calculator to find the solution of the system. ___________

(c) Interpret this solution.

2.4 Lecture Guide:
Solving Linear Equations in One Variable Using the Addition-Subtraction Principle

Objective 1: Solve linear equations in one variable using the addition-subtraction principle.

Linear Equation in One Variable

Algebraically	**Verbally**	**Algebraic Example**
A linear equation in one variable x is an equation that can be written in the form $Ax = B$, where A and B are real constants and $A \neq 0$.	A linear equation in one variable is ________ degree in this variable.	$2x = 24$

1. Which of the following choices are linear equations in one variable?

(a) $3x+1=10$ **(b)** $3x+y=10$ **(c)** $3x^2+1=10$ **(d)** $3x+1-10$

Addition-Subtraction Principle of Equality

Verbally	**Algebraically**	**Numerical Example**
If the same number is added to or subtracted from both sides of an equation, the result is an ______________ equation.	If a, b, and c are real numbers, then $a=b$ is equivalent to $a+c=$ ______ and to $a-c=$ ______ .	$x+3=7$ is equivalent to $x+3-3=7-$ ______ and to $x=$ ______

Use the addition-subtraction principle of equality to solve each equation. Note that we can check our solution of each equation.

2. $x-2=-5$

3. $x+3=3$

4. $5m=4m-3$

5. $7d=6d$

6. $4x-5=3x+8$

7. $8x-2=7x+12$

8. $3x+6=2x+6$

9. $5x-8=2(2x+3)$

10. $7(2x-2)=13(x+1)$

11. $x-10+5x=8x+6-3x$

Based on the limited variety of equations we have examined, a good strategy to solve a linear equation in one variable is:

1. Use the ________________ property to remove any ________________ symbols.
2. Use the addition-subtraction principle of equality to move all ________________ terms to one side.
3. Use the addition-subtraction principle of equality to move all ________________ terms to the other side.

Objective 2: Use graphs and tables to solve a linear equation in one variable.

The solution of a linear equation in one variable is an x-value that causes both sides of the equation to have the same value. To solve a linear equation in one variable using graphs or tables, let Y_1 equal the left side of the equation and let Y_2 equal the right side of the equation. Using a graph, look for the ______-coordinate of the point of __________________ of the two graphs. Using a table of values, look for the ______-value where the two ______-values are equal. Note that the solution of a linear equation in one variable is an x-value and not an ordered pair.

12. Use the graph shown to determine the solution of the equation $2x-1=x+2$.

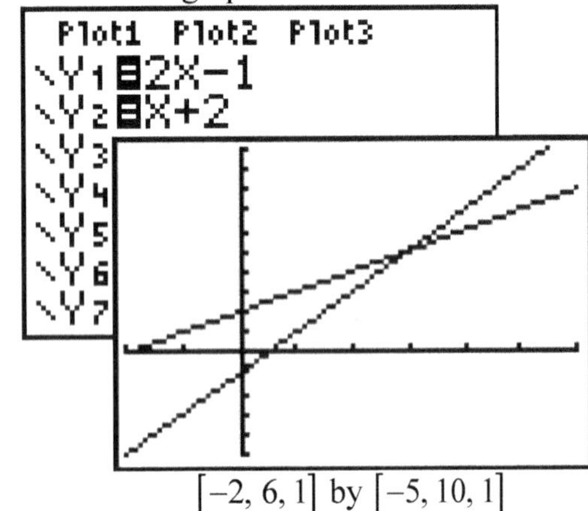

$[-2, 6, 1]$ by $[-5, 10, 1]$

The point where the two lines intersect has an x-coordinate of ______.

Solution: __________________

Verify your result by solving $2x-1=x+2$ algebraically.

13. Use the table shown to determine the solution of the equation $3x-1=2x+6$.

Plot1 Plot2 Plot3
\Y1=3X-1
\Y2=2X+6
\Y3=
\Y4=
\Y5=
\Y6=
\Y7=

X	Y1	Y2
4	11	14
5	14	16
6	17	18
7	20	20
8	23	22
9	26	24
10	29	26

X=4

The x-value in the table at which the two y values are equal is ______.

Solution: __________________

Verify your result by solving $3x-1=2x+6$ algebraically.

Solve each equation using a table or a graph from your calculator by letting Y_1 equal the left side of the equation and Y_2 equal the right side of the equation. See Technology Perspective 2.3.2 for help.

14. $0.5x+3=-0.5x-5$

Solution: ____________

15. $2(x+3)-3x=5-2x$

Solution: ____________

Once the viewing window has been adjusted so you can see the point of intersection of two lines, the keystrokes required to find that point of intersection are ______ ______ ______ ______ ______ ______. To view a graph in the standard viewing window, press **ZOOM** ______.

Objective 3: Identify a linear equation as a conditional equation, an identity, or a contradiction.

There are three classifications of linear equations: **conditional equations, identities, and contradictions.** Each of the equations in problems 2-11 is called a ________________ ________________ because it is only true for certain values of the variable and untrue for other values.

Conditional Equation:
A conditional equation is true for some values of the variable and false for other values.

Algebraic Example
$2x = x + 3$

Solution: $x = 3$

The only value of x that checks is $x = 3$.

Graphical Example

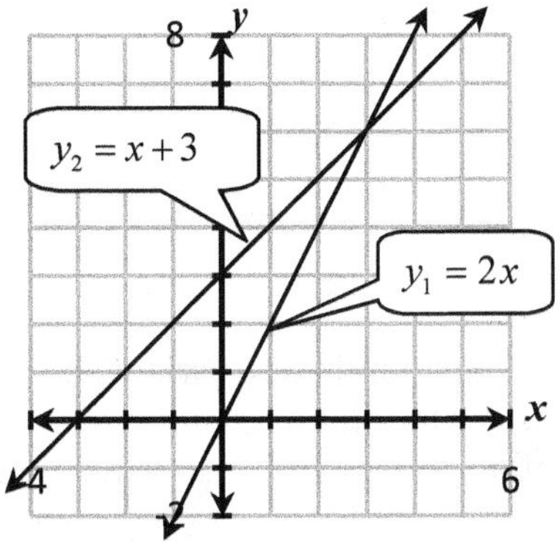

The lines intersect at an x-value of 3.

Numerical Example

x	$y_1 = 2x$	$y_2 = x + 3$
−1	−2	2
0	0	3
1	2	4
2	4	5
3	6	6
4	8	7

The table values of $y_1 = 2x$ and $y_2 = x + 3$ are equal for $x = 3$.

Identity:
An identity is an equation that is true for all values of the variable.

Algebraic Example
$2x = x + x$

Solution: All real numbers.

All real numbers will check. $x + x$ is always $2x$.

Graphical Example

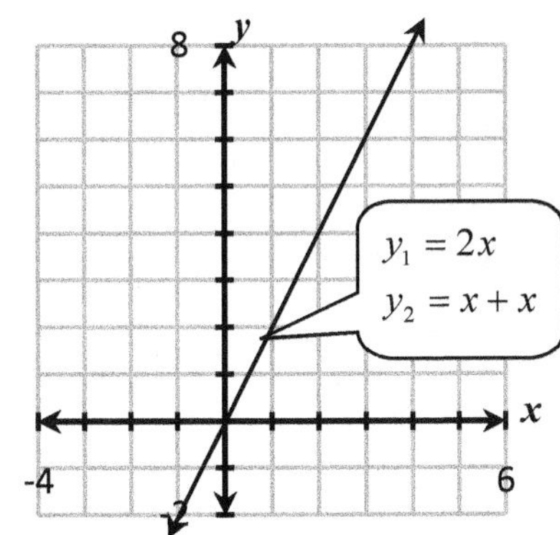

You see only one line because the lines coincide for all values of x.

Numerical Example

x	$y_1 = 2x$	$y_2 = x + x$
−1	−2	−2
0	0	0
1	2	2
2	4	4
3	6	6
4	8	8

The table values of $y_1 = 2x$ and $y_2 = x + x$ are equal for all values of x.

Contradiction:
A contradiction is an equation that is false for all values of the variable.

Algebraic Example

$x = x + 3$

Solution: No solution.

No real numbers will check because no real number is 3 greater than its own value.

Graphical Example

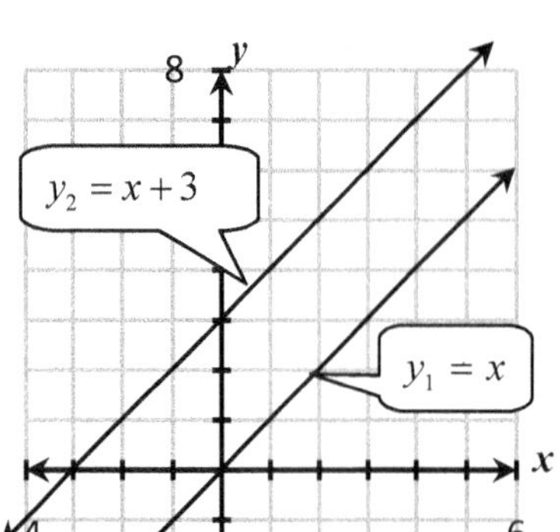

These lines have no points in common.

Numerical Example

x	$y_1 = x$	$y_2 = x + 3$
−1	−1	2
0	0	3
1	1	4
2	2	5
3	3	6
4	4	7

The table values of $y_1 = x$ and $y_2 = x + 3$ will never be equal for any value of x.

Tip: If the solution process for solving a linear equation in one variable produces a unique solution, then the original equation is a conditional equation. If the solution process results in the variable disappearing from both sides of the equation, then the equation you are trying to solve is either a contradiction or an identity

Identify each equation as a contradiction or identity and write the solution of the equation.

16. $3x + 2 - x = 4 + 2x$

17. $5x + 3 - 2x = 3x + 3$

18. $4(2x - 3) = 7x - 12 + x$

19. $3x + 4 + 2x = 5(x - 1)$

Simplify vs Solve

Simplify the expression in the first column by combining like terms, and solve the equation in the second column.

20. Simplify $7x+3+6x-5$

21. Solve $7x+3=6x-5$

22. Simplify $3x+4-(2x-8)$

23. Solve $3x+4=2x-8$

Translate each verbal statement into algebraic form.

24. Eight more than three times a number is equal to three less than the number.

25. Seven less than six times a number equals two times the quantity of eight less than a number.

26. Write an algebraic equation for the following statement, using the variable m to represent the number, and then solve for m.

Verbal Statement: Five less than three times a number is equal to two times the sum of the number and three.

Algebraic Equation:

Solve this equation:

27. The perimeter of the parallelogram shown equals $(26 + a)$ cm. Find a.

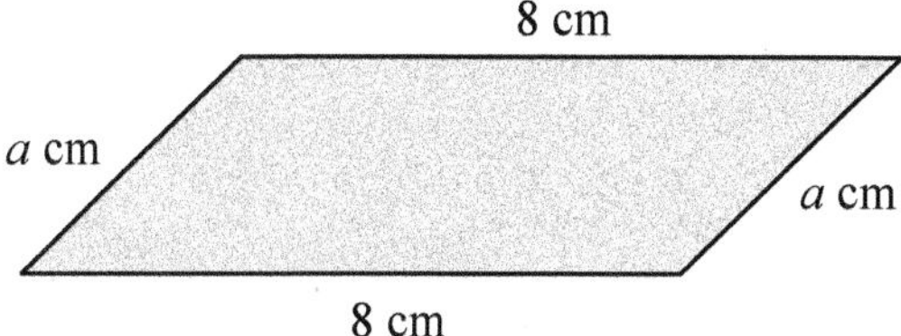

2.5 Lecture Guide: Solving Linear Equations in One Variable Using the Multiplication-Division Principle

Objective 1: Solve linear equations in one variable using the multiplication-division principle.

Multiplication-Division Principle of Equality

Verbally	**Algebraically**	**Numerical Example**
If both sides of an equation are multiplied or divided by the same nonzero number, the result is an ____________ equation.	If a, b, and c are real numbers and $c \neq 0$, then $a = b$ is equivalent to $ac =$ _____ and to $\frac{a}{c} =$ _____.	$\frac{x}{2} = 5$ is equivalent to $2\left(\frac{x}{2}\right) = 2(5)$; and $3x = 12$ is equivalent to $\frac{3x}{3} = \frac{12}{3}$.

Solve each equation.

1. $8x = 72$

2. $\frac{x}{3} = 7$

Strategy for Solving Linear Equations

Step 1. ____________ each side of the equation.

a. If the equation contains fractions, simplify by ____________ both sides of the equation by the least common denominator (LCD) of all the fractions.

b. If the equation contains grouping symbols, simplify by using the distributive property to remove the grouping symbols and then ____________ like terms.

Step 2. Using the addition-subtraction principle of equality, isolate the variable terms on one side of the equation and the ____________ terms on the other side.

Step 3. Using the multiplication-division principle of equality, solve the equation produced in Step 2.

The following examples require using the multiplication-division principle of equality. Some will also require using the addition-subtraction principle of equality. Solve each equation. Note that we can check our solutions of each equation.

3. $x - 2 = 6$

4. $x + 2 = 6$

5. $\frac{x}{2}=6$

6. $2x=6$

7. $-a=12$

8. $\frac{x}{5}=-4$

9. $7x+5=-23$

10. $8t+1=-3t+23$

11. $2(3x-1)=2x+22$

12. $3(4x-5)=5(2x+1)$

13. $5-3(2x+1)=4(2-x)$

14. $\frac{x}{3}-1=\frac{x}{2}+4$

15. $\frac{m+1}{5}-3=\frac{m+7}{12}$

16. $\frac{5x+1}{3}=1-\frac{3x-2}{5}$

17. $6(x-1)-3(x+2)=4(2x+3)-(3x-4)$

18. Note the difference between simplifying expressions and solving equations:

(a) Simplify $3(5x+2)-2(8x+1)$ **(b) Solve** $3(5x+2)=2(8x+1)$

Translate each verbal statement into algebraic form.

19. Twice the sum of a number and 3 is equal to eleven less than four times the number.

20. One-half the quantity of a number plus three is the same as five minus the number.

21. Write an algebraic equation for the following statement and then solve the equation.

Verbal Statement: Three times the quantity of a number plus four is two less than the number.

Algebraic Equation:

Solve this equation:

22. The perimeter of the rectangle shown equals 44 cm. Find *a*.

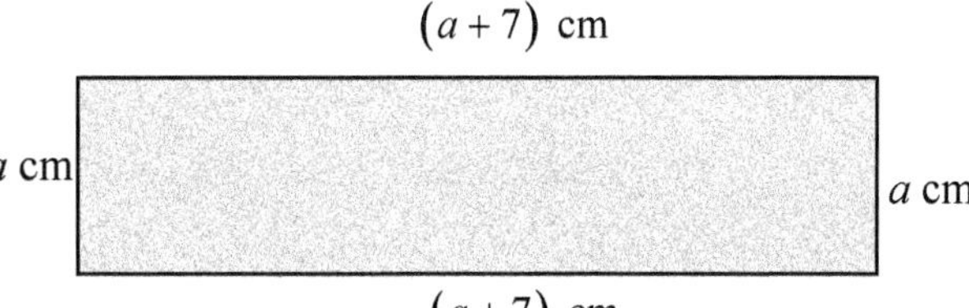

23. Solve the equation $x-4=-2x+5$ by letting Y_1 equal the left side of the equation and Y_2 equal the right side of the equation.

(a) Use your calculator to create a graph of Y_1 and Y_2 using a viewing window of $[-5, 5, 1]$ by $[-5, 5, 1]$. Use the **Intersect** feature to find the point where these two lines intersect. Draw a rough sketch below. The values in the table will help.

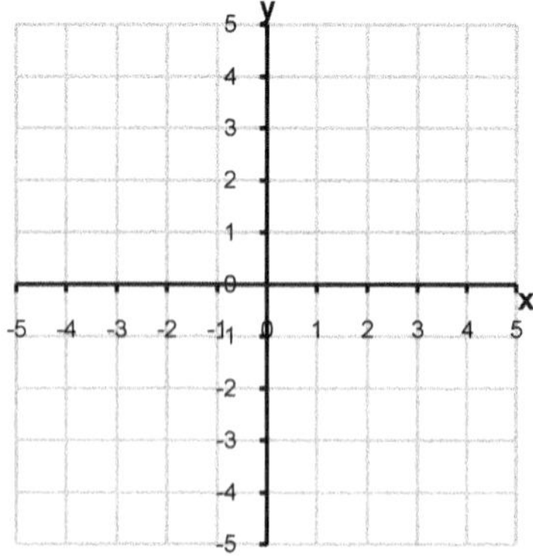

The point where the two lines intersect has an x-coordinate of ______.

(b) Create a table on your calculator with the table settings: TblStart = 0; ΔTbl = 1
Complete the table below.

x	Y_1	Y_2
0		
1		
2		
3		
4		
5		
6		

The x-value at which the two y values are equal is ______.

(c) Solve the equation $x-4=-2x+5$ algebraically.

(d) Check your solution.

2.6 Lecture Guide: Calculating Intercepts and Rearranging Formulas

To solve a linear equation for a specified variable, we want to isolate that variable on one side of the equation with all other variables on the other side of the equation. This process involves the same steps used to solve a linear equation in one variable. If there is more than one variable in the equation, it is helpful to think of all variables as constants except the variable you are solving for.

Objective 1: Rewrite a linear equation in the form $y = mx + b$.

Solve each equation below for *y*, and write it in the form $y = mx + b$.

1. $3x + 2y = 12$

2. $5x - 6y = 30$

3. $8x - 2y = 10$

4. $\frac{x}{3} + \frac{y}{4} = 1$

5. $1.5x - 0.3y = 1.2$

6. $3(x - 2y) = 2(5x + 3y - 12)$

Objective 2: Calculate the x- and y-intercepts of a line.

Algebraically finding the x- and y-intercepts

To calculate the x-intercept: Substitute ______ for y and solve for x.	To calculate the y-intercept: Substitute ______ for x and solve for y.

Calculate the x- and y-intercepts of each line, and then use the intercepts to graph the line.

7. $3x - 4y = 12$

Graph:

x-intercept: **y-intercept:**

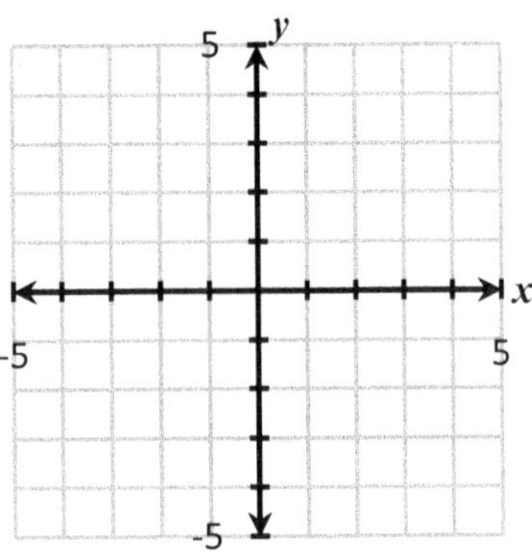

8. $y = \frac{2}{3}x - 4$

Graph:

x-intercept: **y-intercept:**

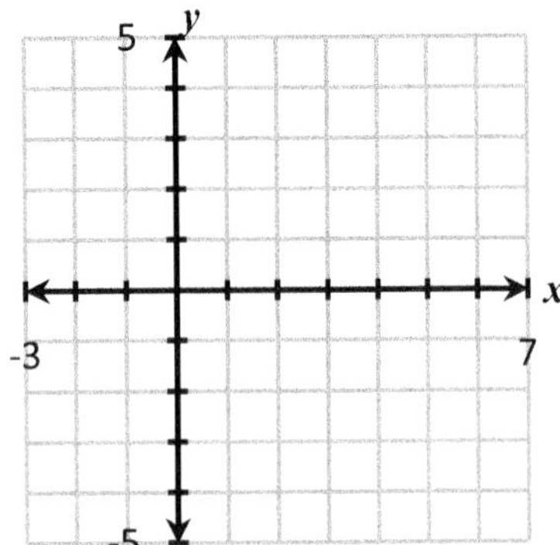

9. A young entrepreneur operating a gum ball machine has a fixed monthly overhead cost of \$5. He makes a profit of \$0.20 on each gum ball sold.

(a) Write an equation that gives the monthly profit for this business when x gum balls are sold in a month.

(b) Using the equation from part (a), determine the x-intercept of the graph of this equation and interpret this point.

(c) Using the equation from part (a), determine the y-intercept of the graph of this equation and interpret this point.

Objective 3: Solve an equation for a specified variable.
Solve each equation for the specified variable.

10. $A = \frac{1}{2}bh$ for b

11. $P_1V_1 = P_2V_2$ for P_1

12. $P = 2l + 2w$ for l

13. $A = \frac{1}{2}h(a+b)$ for a

14. $y = mx + b$ for x

15. $2m + 4t = 2 - a - 6m$ for m

2.7 Lecture Guide: Proportions and Direct Variation

Objective 1: Solve problems involving proportions.

Proportion

Algebraically	**Verbally**	**Numerical Example**
$\frac{a}{b}=\frac{c}{d}$ or $a:b=c:d$	This proportion is read "*a* is to *b* as *c* is to *d*." The extremes are *a* and *d*, and the ____________ are *b* and *c*.	$\frac{2}{5}=\frac{40}{100}$ or $2:5=40:100$

Solve the following proportions.

1. $\frac{x}{4}=\frac{3}{2}$

2. $\frac{x}{12}=\frac{4}{3}$

3. $\frac{2x}{3}=\frac{4}{15}$

4. $\frac{15}{x}=\frac{5}{4}$

5. $x:7=9:70$

6. $3:8=9:4x$

7. $\frac{6x-10}{5}=\frac{4x+4}{2}$

8. $\frac{2y-3}{3}=\frac{6y+5}{7}$

9. A quality-control inspector found 5 defective computer chips in the 2500 that were tested. How many defective chips would be expected in a shipment of 100,000 chips?

10. The FDA allows 13 insect heads for every 100 grams of fig paste used in fig cookies. How many insect heads would be allowed in 5000 grams of fig paste?

11. On a drawing of a set of house plans 3 cm represents a distance of 15 ft. What distance corresponds to 7.5 cm on the house plans?

Objective 2: Solve problems involving direct variation.

Direct Variation

If x and y are real variables and k is a real constant, then:

Verbally	**Algebraically**	**Numerical Example**	**Graphical Example**
y varies directly as x with_____________ of variation k.	$y = kx$ *Example:* $y = 3x$	see table below	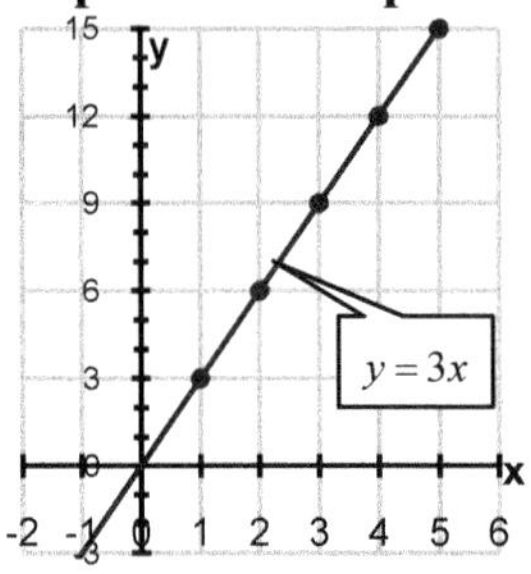 $y = 3x$

x	$y = 3x$
1	3
2	6
3	9
4	12
5	15

12. If y varies directly as x and y is 3 when x is 4, find y when x is 24.

13. If m varies directly as n and m is 12 when n is 5, find m when n is 60.

14. The distance a plane travels varies directly with time. The plane travels 1275 miles in 3 hrs.

(a) What is the rate at which the plane travels in mi/h? (This is the constant of variation.)

(b) Use the rate to write an equation relating the distance the plane travels to the time it takes to travel that distance.

(c) Use this equation and a graphing calculator to complete the table of values shown below. Write the values from your table here.

Time, x	Distance, Y_1
0	
1.5	
3	
4.5	
6	
7.5	
9	

2.8 Lecture Guide: More Applications of Linear Equations

Objective 1: Determine the restrictions on a variable in an application.

1. The function $C(x) = 85x$ models the cost of renting an auger for x days in one week. The rental company will charge for a full day even if the auger is used for a partial day. What restrictions should be placed on the variable x?

2. For two investments totaling \$3000, the function $f(x) = 3000 - x$ models the amount of a second investment where x is the amount of the first investment. What restrictions should be placed on the variable x?

3. You have 40 feet of fencing to enclose three sides of a rectangular pen against a wall.

(a) If x represents the amount of fencing used for the width of the pen, write a function $f(x)$ that represents the amount of fencing remaining for the length of the pen.

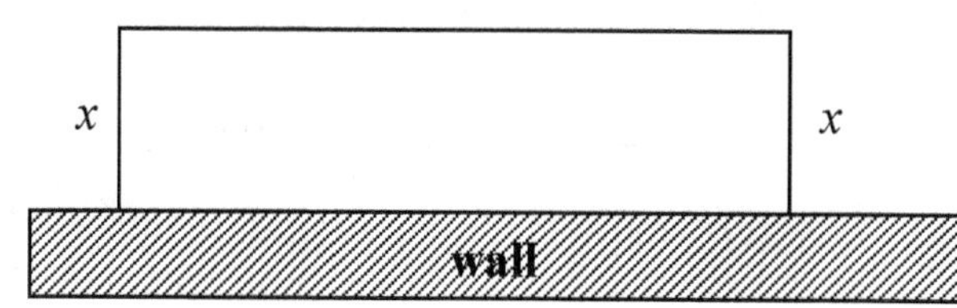

(b) What restrictions should be placed on the variable x?

4. A 12-foot board is to be cut in two pieces.

(a) If x represents the length of one of the pieces, write a function $f(x)$ that represents the length of the other piece.

(b) What restrictions should be placed on the variable x?

Strategy for Solving Word Problems

Step 1. Read the problem carefully to determine what you are being asked to find.
Step 2. Select a ____________ to represent each unknown quantity. Specify precisely what each variable represents and note any restrictions on each variable.
Step 3. If necessary, make a sketch and translate the problem into a word equation or a system of word equations. Then translate each word equation into an ____________ equation.
Step 4. Solve the equation or system of equations, and answer the question completely in the form of a sentence.
Step 5. Check the ________________ of your answer.

Objective 2: Use the mixture principle to create a mathematical model.

Mixture Principle for Two Ingredients

Amount in first + Amount in second = Amount in mixture

Applications of the Mixture Principle

1. Amount of product A + Amount of product B = Total amount of mixture
2. Variable cost + Fixed cost = Total cost
3. Interest on bonds + Interest on CDs = Total interest
4. Distance by first plane + Distance by second plane = Total distance
5. Antifreeze in first solution + Antifreeze in second solution = Total amount of antifreeze

5. A local cell phone provider charges a fixed fee of \$35 per month plus a variable charge of \$0.23 per minute. If a customer received a phone bill for \$90.20, how many cell phone minutes were used during the month of service?

(a) Identify the variable:

Let m = the number of ______________ used for the ____________

(b) Write the word equation:

____________ cost + fixed ______________ = total cost

(c) Translate the word equation into an algebraic equation:

__________ + __________ = 90.20

(d) Solve this equation:

(e) Write a sentence that answers the question:

(f) Is this answer reasonable?

6. **Interest from Two Investments**
A broker made two separate investments earning a total of $1100 in interest. The first investment was for $10,000 and earned 8%. If the second investment earned 5%, determine the amount of the second investment.

(a) Identify the variable:

Let x = number of ____________ in the ____________ investment

(b) Write the word equation:

__________ from the first investment + __________ from the the second investment = total ______________

(c) Translate the word equation into an algebraic equation:

______________ + ____________ = 1100

(d) Solve this equation:

(e) Write a sentence that answers the question:

(f) Is this answer reasonable?

7. **Travel Time** From a point on a straight road, John and Fred ride bicycles in opposite directions. John rides 10 miles per hour and Fred rides 2 miles per hour faster than John. With continuous riding, how many hours will it take them to travel 55 miles apart? Neither can cycle for more than 3 hours without taking a break.

   (a) Identify the variable:

   Let t = time in ________ to travel _________ miles apart

   (b) Write the word equation:

   _____________ John travels + ________________ Fred travels = Total ________________

   (c) Translate the word equation into an algebraic equation:

   ____________ + _____________ = 55

   (d) Solve this equation:

   (e) Write a sentence that answers the question:

   (f) Is this answer reasonable?

8. **Mixture of Milk** In a 175 liter container, a farmer has 100 liters of milk that is 4.6% butterfat. One customer requests milk that is 3.2% butterfat. How much skim milk (no butterfat) should be added to the 4.6% butterfat to make milk that is 3.2% butterfat?

(a) Identify the variable:

Let $x =$ number of __________ of ____________ milk to add to the container

Let __________ = number of __________ of milk in the mixture

(b) Write the word equation:

Butterfat in 100L of 4.6% milk	+	____________ in the skim milk	=	total butterfat in the ______________

(c) Translate the word equation into an algebraic equation:

_________________ + _________________ = _________________

(d) Solve this equation:

(e) Write a sentence that answers the question:

(f) Is this answer reasonable?

9. How would problem 8 be different if the farmer had started with a 125 liter container?

Lecture Guides for

Chapter 3

Lines and Systems of Linear Equations in Two Variables

3.1 Lecture Guide: Slope of a Line and Applications of Slope

Objective 1: Determine the slope of a line.

Slope of a Line Through (x_1, y_1) and (x_2, y_2)

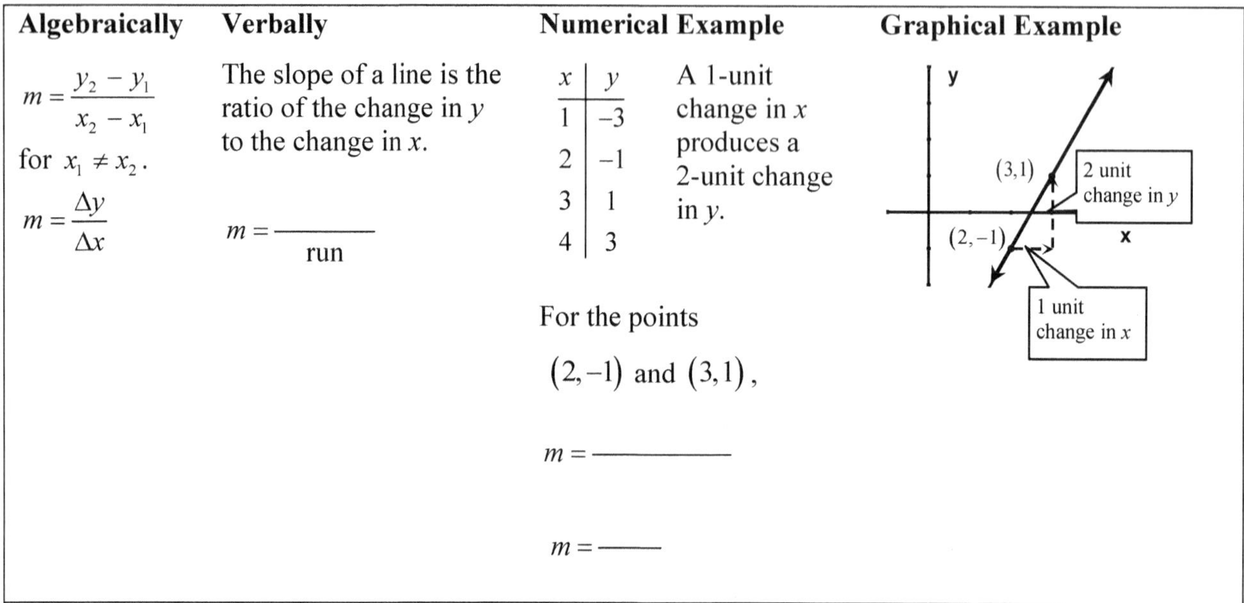

Algebraically	Verbally	Numerical Example	Graphical Example
$m = \dfrac{y_2 - y_1}{x_2 - x_1}$ for $x_1 \neq x_2$. $m = \dfrac{\Delta y}{\Delta x}$	The slope of a line is the ratio of the change in y to the change in x. $m = \dfrac{\quad}{\text{run}}$	x: 1, 2, 3, 4; y: −3, −1, 1, 3. A 1-unit change in x produces a 2-unit change in y. For the points $(2,-1)$ and $(3,1)$, $m = \underline{\qquad}$ $m = \underline{\quad}$	

Calculate the slope of the line through each pair of points then graph a line that passes through the points.

1. $(-2,7)$ and $(3,5)$

2. $(1,-8)$ and $(7,-3)$

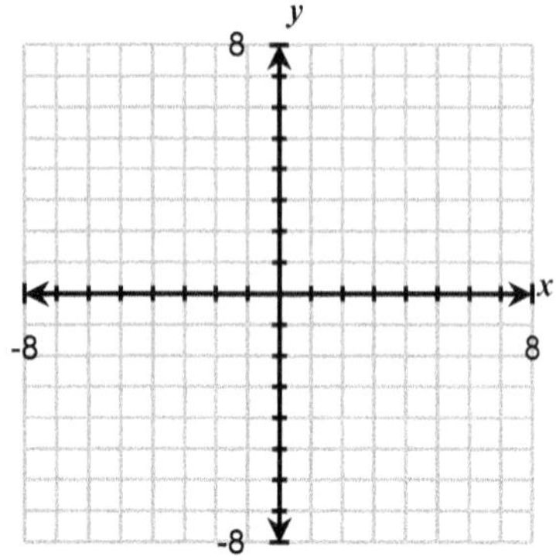

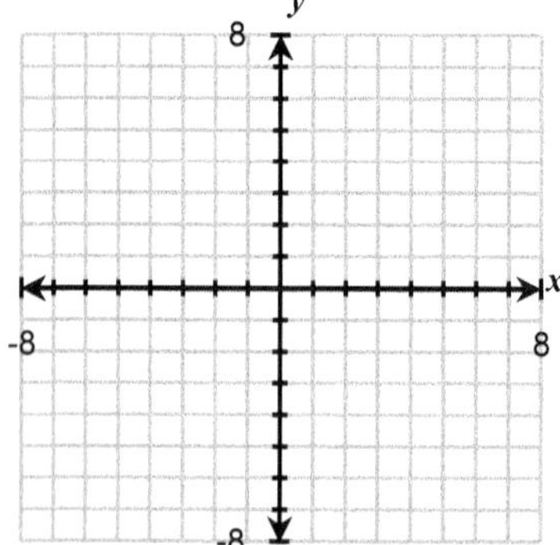

3. $(-5,3)$ and $(2,3)$

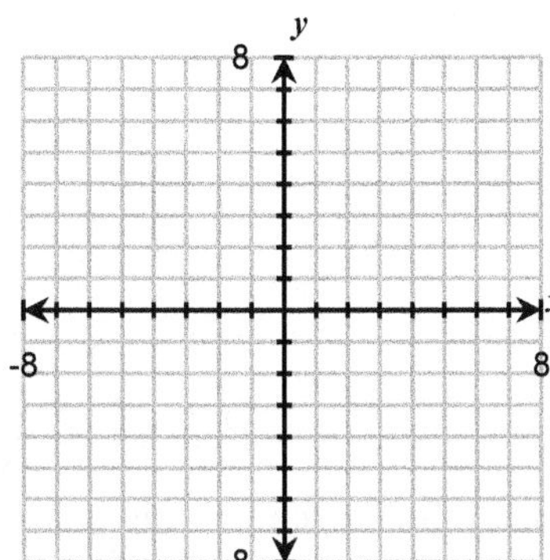

4. $(5,3)$ and $(5,-2)$

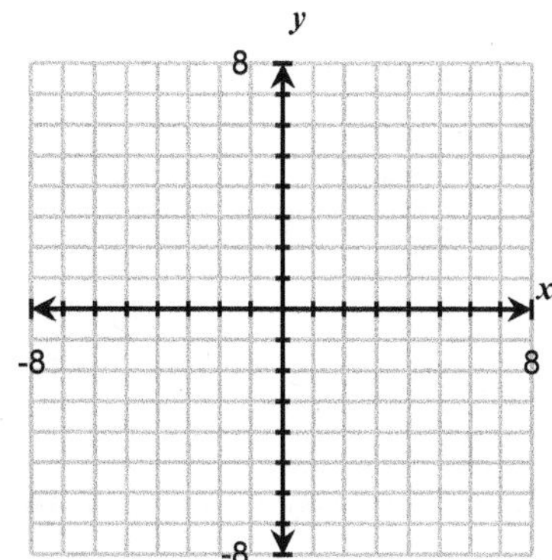

Classifying Lines by Their Slopes

Numerically	**Verbally**
m is positive	The line slopes ______________________ to the right.
m is negative	The line slopes ______________________ to the right.
m is zero	The line is ______________________.
m is undefined	The line is ______________________.

5. Calculate the slope of the line in the graph.

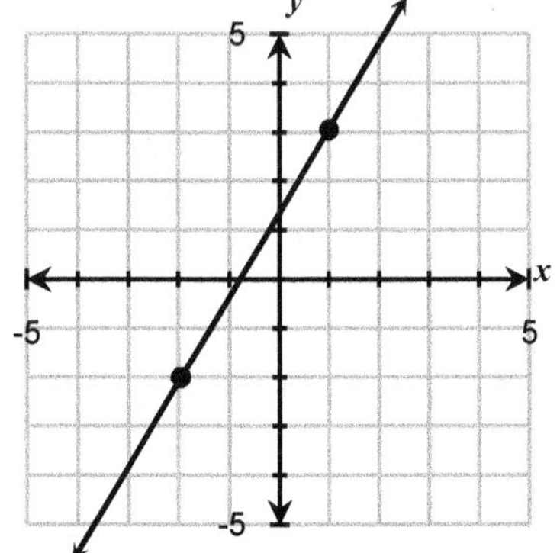

6. Calculate the slope of the line in the graph.

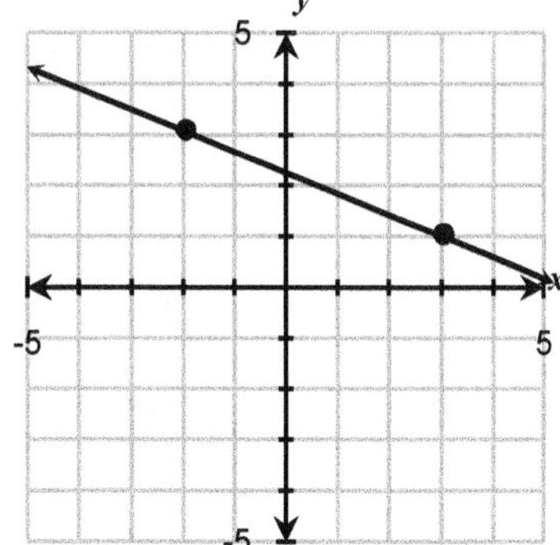

7. Determine the slope of the line in the graph.

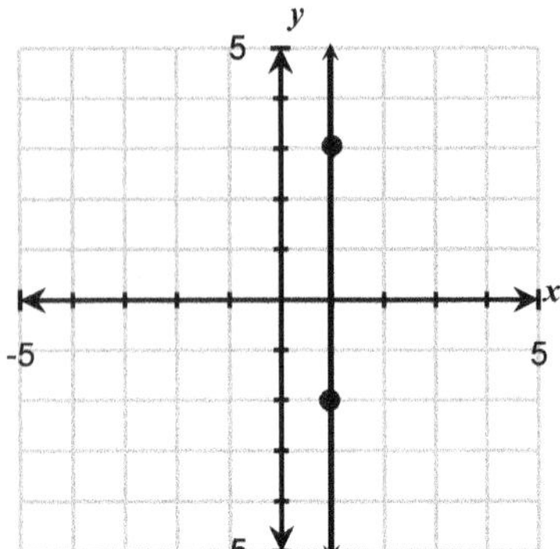

8. Determine the slope of the line in the graph.

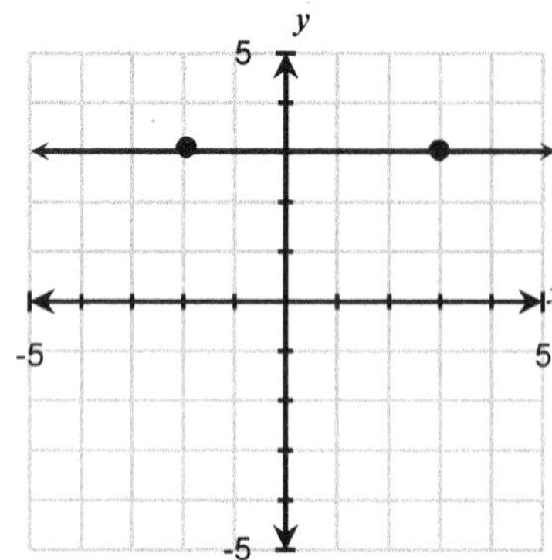

9. Calculate the slope of the line containing the points in the table.

x	y
0	2
5	5
10	8
15	11
20	14
25	17
30	20

10. Calculate the slope of the line containing the points in the table.

x	y
−3	4
0	2
3	0
6	−2
9	−4
12	−6
15	−8

11. Complete the table so that the points all lie on a line having a slope $m = \frac{5}{4}$.

x	y
0	3
4	
8	
12	
16	

12. Complete the table so that the points all lie on a line having a slope $m = -\frac{3}{4}$.

x	y
0	3
4	
8	
12	
16	

13. For the equation $4x-6y=-24$

(a) Find the x-intercept.

(b) Find the y-intercept.

(c) Use the points to determine the slope of the line.

Objective2: Use slopes to determine whether two lines are parallel, perpendicular, or neither.

Parallel and Perpendicular Lines

If l_1 and l_2 are distinct nonvertical* lines with slopes m_1 and m_2 respectively, then:

Algebraically	**Verbally**	**Graphically** 
$m_1 = m_2$	l_1 and l_2 are parallel because they have the ____________ slope.	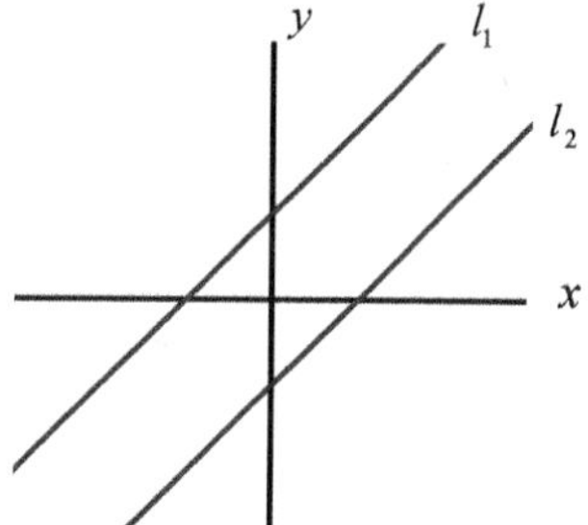
$m_1 = -\dfrac{1}{m_2}$ or $m_1 m_2 = -1$	l_1 and l_2 are perpendicular because their slopes are negative ____________.	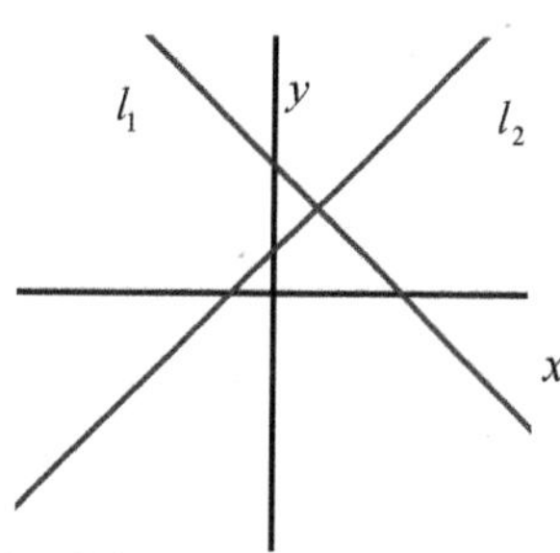

* Also note all vertical lines are parallel to each other, and all vertical lines are perpendicular to all horizontal lines.

14. a. If l_1 and l_2 are parallel and $m_1 = \frac{3}{5}$, then $m_2 =$ ______.

b. If l_1 and l_2 are perpendicular and $m_1 = \frac{3}{5}$, then $m_2 =$ ______.

15. If l_1 and l_2 are perpendicular and $m_1 = -4$, then $m_2 =$ ______.

16. If l_1 and l_2 are perpendicular and $m_1 = 0$, then m_2 is ______________.

Determine whether the line that passes through the first pair of points is parallel to, perpendicular to, or neither parallel nor perpendicular to the line that passes through the second pair of points.

17. $(-2,1)$ and $(3,5)$
$(6,-2)$ and $(2,3)$

18. $(-1,2)$ and $(-3,5)$
$(4,0)$ and $(2,-3)$

19. $(-2,5)$ and $(0,1)$
$(7,3)$ and $(6,5)$

20. $(-3,4)$ and $(1,7)$
$(0,-6)$ and $(3,-2)$

21. $(-3,4)$ and $(6,4)$
$(-2,5)$ and $(-2,0)$

22. $(-3,4)$ and $(6,4)$
$(-2,1)$ and $(3,1)$

23. Compute the missing values in the table.

Change in x	**Change in y**	**Slope**
−5	3	
7		$-\frac{1}{7}$
	4	2
3	0	
	3	Undefined

Using the given point and slope, determine another point on the line and graph the line.

24. Through $(0,-3)$ with $m=\frac{1}{2}$

Point: ________

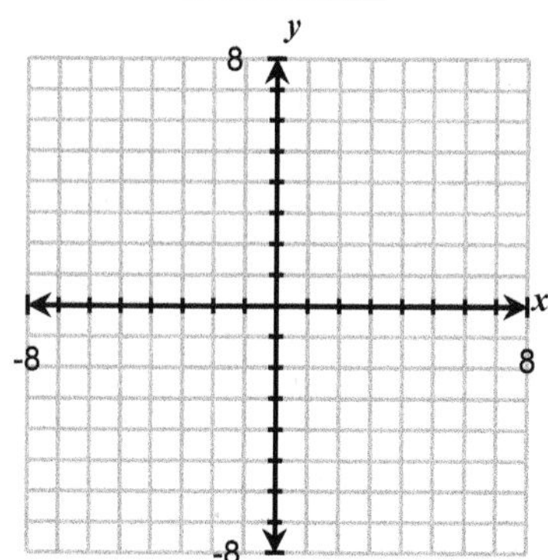

25. Through $(0,2)$ with $m=-\frac{2}{3}$

Point: ________

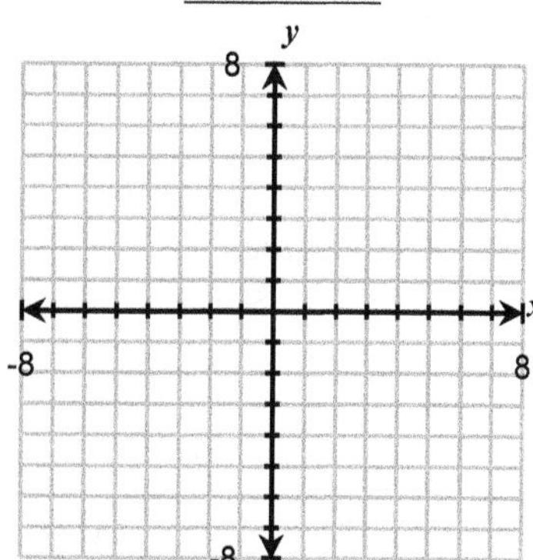

26. Through $(4,-3)$ with $m=0$

Point: ________

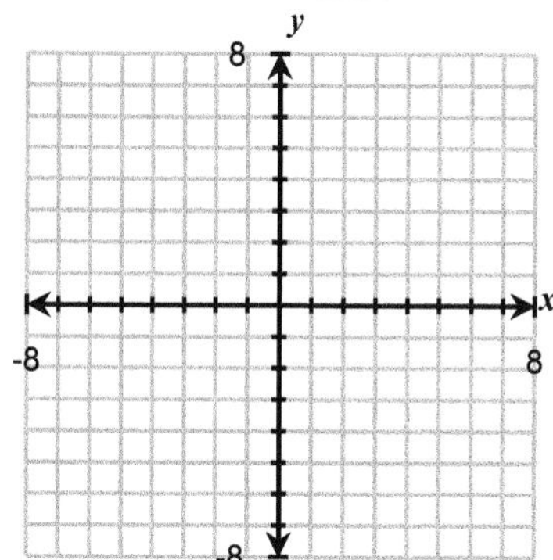

27. Through $(4,-3)$ with an undefined slope.

Point: ________

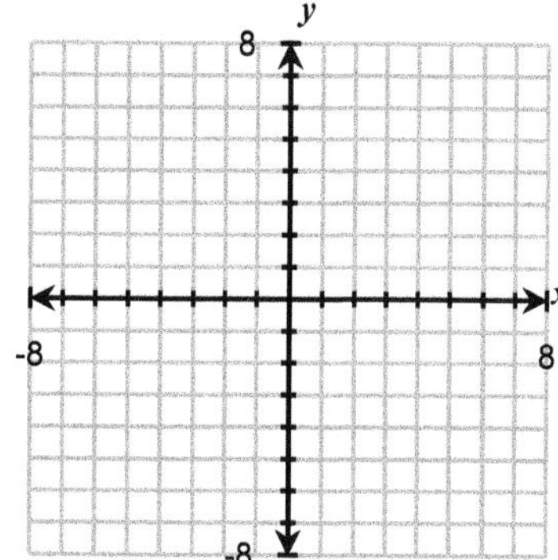

Objective 3: Calculate and interpret rates of change.

28. A local high school purchases a copy machine for $1200. Due to depreciation, the value of the machine decreases with time. The table below lists the value y of the copy machine after x months.

Months	Value
0	$1200
6	$1050
12	$900
18	$750
24	$600
30	$450
36	$300

(a) Determine the rate of change of the value with respect to time.

(b) Interpret the meaning of this value.

(c) At this rate, how long after the copy machine was purchased will the machine have no value?

3.2 Lecture Guide: Special Forms of Linear Equations in Two Variables

***Objective 1:** Use the slope-intercept form to write and graph linear equations.*

Slope-Intercept Form

Algebraically	**Algebraic Example**	**Verbal Example**	**Graphical Example**
$y = mx + b$ is the equation of a line with ____________ m and y-intercept __________.	$y = \frac{1}{2}x + 3$	This line has slope $\frac{1}{2}$ and a y-intercept of $(0, 3)$.	(2,4) (0,3) 1 2 $y = \frac{1}{2}x + 3$

1. The line $y = \frac{2}{5}x - 4$ has a slope of __________ and a y-intercept of ____________.

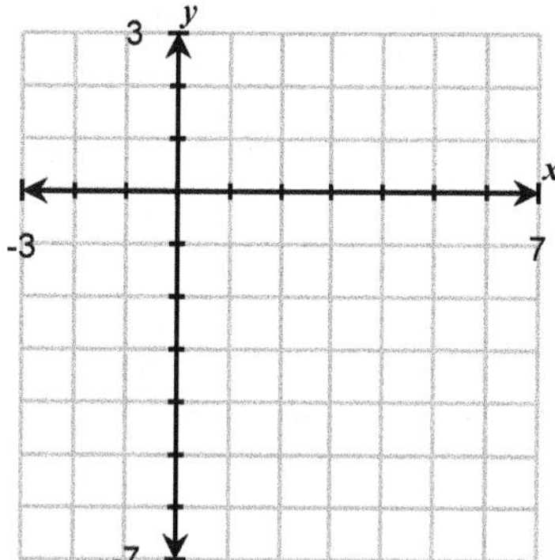

2. Graph the line $y = \frac{2}{5}x - 4$ using the slope and y-intercept.

3. Complete the following table. This example stresses the fact that if we know the slope and the y-intercept, then we can immediately write the equation. Also, if we have the equation in slope-intercept form, we can immediately sketch the graph because we can quickly determine the slope and the y-intercept.

Slope	y-intercept	Equation
$\frac{2}{3}$	$(0, 1)$	
-3	$(0, -4)$	
		$y = -\frac{5}{3}x + 7$
		$f(x) = -\frac{1}{3}x + 5$

To graph a line or to give the equation of the line, it is sufficient to know any point on the line and the slope of the line. This is equivalent to knowing any two points on the line because we can calculate the slope given any two points on the line.

Complete the missing information. On all graphs, clearly label at least two points.

4. Equation: $y = \frac{3}{2}x - 2$

Through the point:

Slope:

Graph:

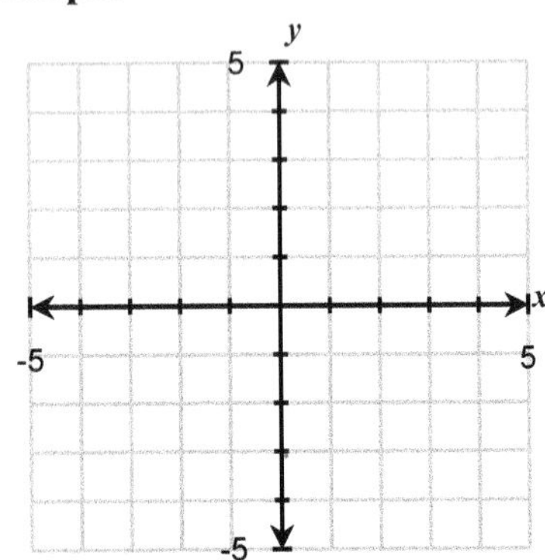

5. Equation: $2y + 3x = 6$

Through the point:

Slope:

Graph:

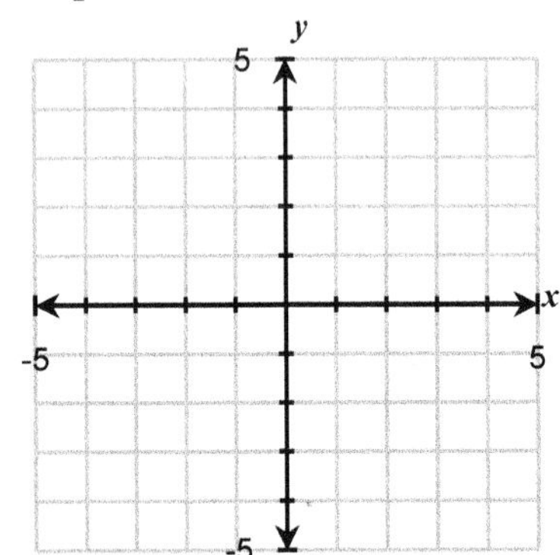

Complete the missing information.

6. Graph:

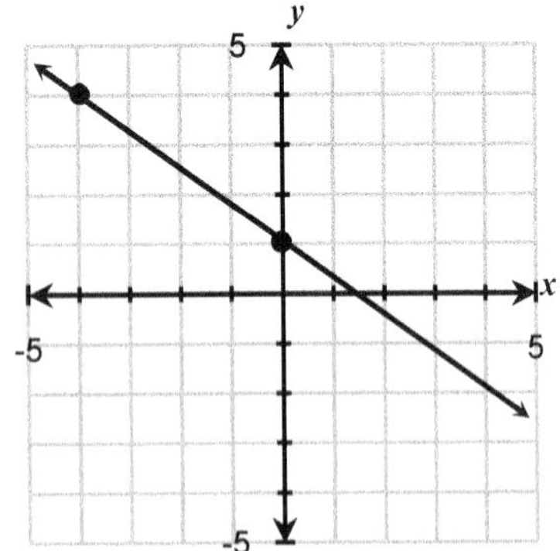

Through the point::

Slope:

Equation:

7. Graph:

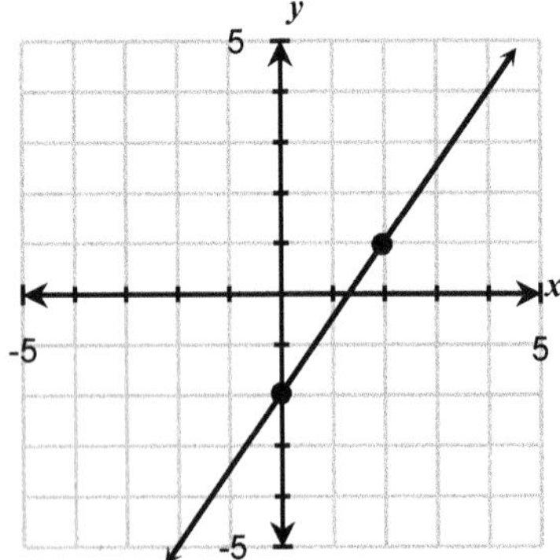

Through the point:

Slope:

Equation:

8. Use the given table of values for a linear function to determine the following.

x	y
−3	4
0	2
3	0
6	−2
9	−4
12	−6
15	−8

(a) The value of Δx.

(b) The value of Δy.

(c) The slope of this line.

(d) The y-intercept of this line.

(e) The equation of this line in slope-intercept form.

Objective 2: Use the point-slope form to write and graph linear equations.

Point-Slope Form $y - y_1 = m(x - x_1)$

Algebraically	**Algebraic Example**	**Graphical Example**	**Verbal Example**
$y - y_1 = m(x - x_1)$ is the equation of a line through __________ with ____________ m.	$y - 1 = \frac{1}{3}(x - 1)$	y; (4,2); (1,1); +1; +3; x	This line passes through the point $(1,1)$ with slope $\frac{1}{3}$.

9. Complete the following table. This example stresses the fact that the point-slope form of a line is useful when the slope and a point other than the y-intercept is given.

Slope	Point	Point-slope equation
4	$(2,1)$	
$\frac{-2}{5}$	$(-3,-4)$	
		$y - 2 = -(x + 3)$
		$y + 4 = \frac{-3}{2}(x + 1)$

Complete the missing information. On all graphs, clearly label at least two points.

10. $y+2=-\frac{1}{2}(x-3)$

Through the point:

Slope:

Graph:

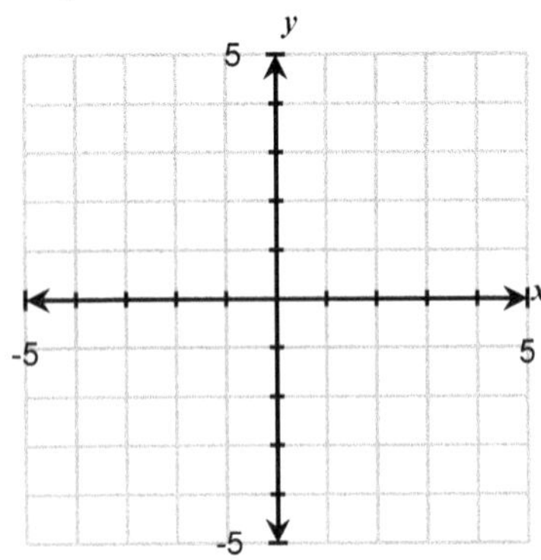

11. $y-3=-2(x+1)$

Through the point:

Slope:

Graph:

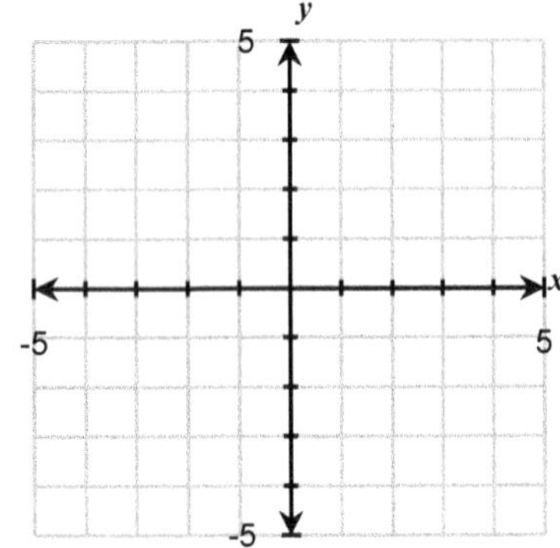

Write each equation in slope-intercept form.

12. $y+2=-\frac{1}{2}(x-3)$

13. $y-3=-\frac{2}{3}(x+1)$

Write the equation of the line passing through the given point with specified slope. Write the answer in slope-intercept form.

14. $(2,-3)$, $m=-3$

15. $(-3,2)$, $m=\frac{1}{2}$

Write the equation of the line passing through the given point with specified slope. Write the answer in slope-intercept form.

16. $(-4, 2)$, $m = -\frac{3}{4}$

17. $(4, -3)$, $m = \frac{2}{3}$

Write in slope-intercept form the equation of the line passing through the given points.

18. $(-2, 4)$ and $(-1, -3)$

19. $(-4, -4)$ and $(1, 2)$

Use the given graph to complete the missing information.

20. Graph:

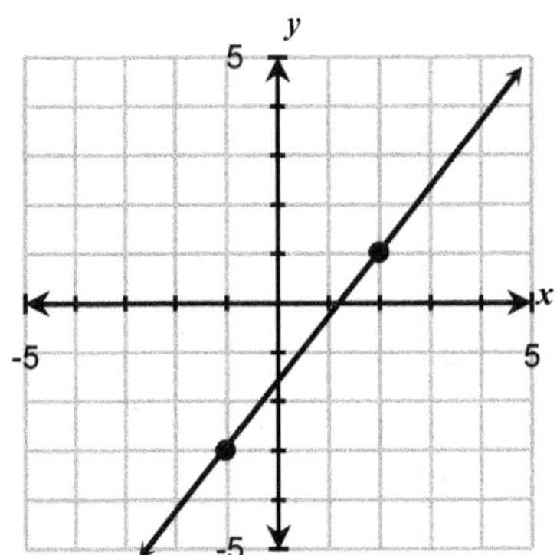

Through the point:

Slope:

Slope-intercept equation:

21. Graph:

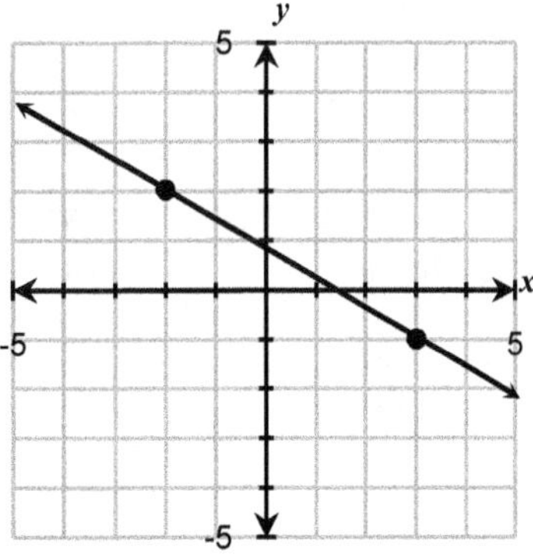

Through the point:

Slope:

Slope-intercept equation:

Objective 3: Use the special forms of equations for horizontal and vertical lines.

Horizontal and Vertical Lines

Algebraically	**Numerical Example**	**Graphical Example**	**Verbally**
$y = b$ is the equation of a ____________ line with y-intercept $(0, b)$. *Example:* $y = 3$	x | y -2 | 3 -1 | 3 0 | 3 1 | 3 2 | 3	$y = 3$	This horizontal line has a y-intercept of $(0, 3)$ and a slope of 0.
$x = a$ is the equation of a ____________ line with x-intercept $(a, 0)$. *Example:* $x = -2$	x | y -2 | -2 -2 | -1 -2 | 0 -2 | 1 -2 | 2	$x = -2$	This vertical line has an x-intercept of $(-2, 0)$ and its slope is undefined.

22. All points on a horizontal line have the same _____-coordinate. This is the reason that the equation of a horizontal line is of the form __________________________. The slope of a horizontal line is _____.

23. All points on a vertical line have the same _____-coordinate. This is the reason that the equation of a vertical line is of the form __________________________. The slope of a vertical line is __________________________.

Graph each equation by completing a table of values and then give any intercepts. Can you check both of these on a graphing calculator? If not, why not?

24. Equation: $x = 3$

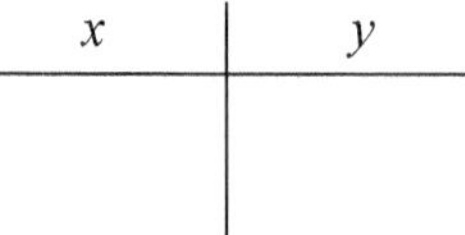

Graph:

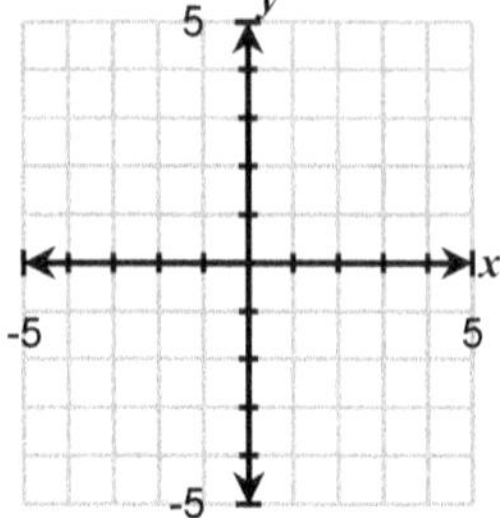

***x*-intercept:** ______

***y*-intercept:** ______

Slope: ______

25. Equation: $y = -2$

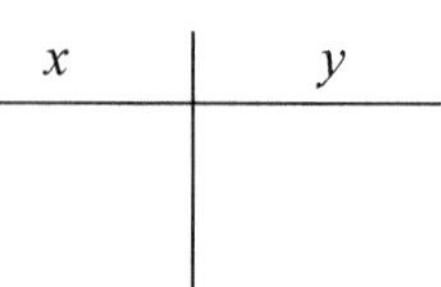

Graph:

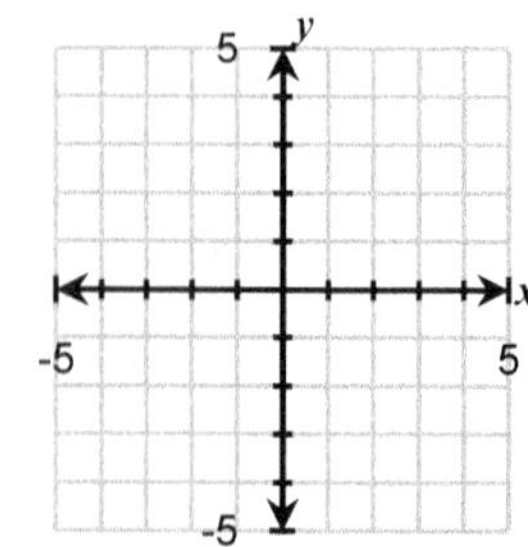

***x*-intercept:** ______

***y*-intercept:** ______

Slope: ______

Complete the missing information. On all graphs, clearly label at least two points.

26. Equation:

Through the point: $(-3, 4)$

Slope: $m = 0$

Graph:

27. Equation:

Through the point: $(4, -3)$

Slope: m is undefined

Graph:

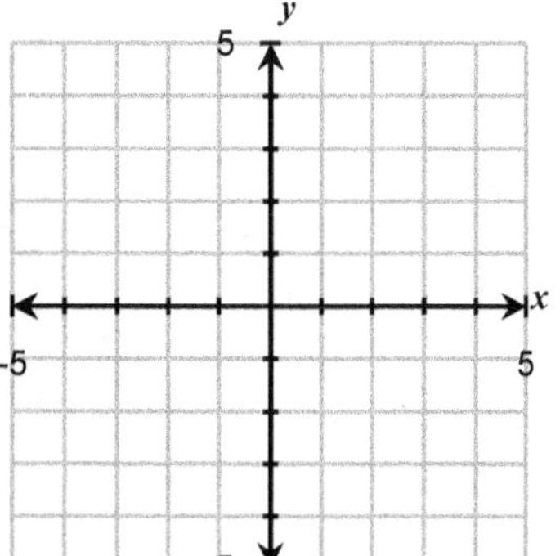

Complete the missing information.

28. Graph:

Through the y-intercept:

Slope:

Equation:

29. Graph:

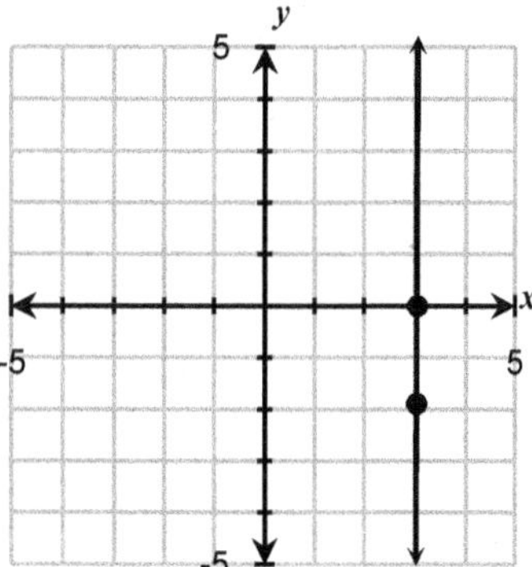

Through the x-intercept:

Slope:

Equation:

Parallel and Perpendicular Lines

Parallel lines have the ____________ slope and _____________________ lines have slopes that are opposite reciprocals.

Determine if the first line is parallel to, perpendicular to, or neither parallel nor perpendicular to the second line.

30. $y = \frac{3}{4}x + 2$

$y = -\frac{3}{4}x - 1$

31. $y = 2x + 3$

$6x + 12y = 24$

32. $3x + 2y = 8$

$6x + 4y = 12$

33. Write in slope-intercept form the equation of a line passing through $(-2, -3)$ and parallel to $y = 5x - 2$.

34. Write in slope-intercept form the equation of a line passing through $(-2, -3)$ and perpendicular to $y = 5x - 2$.

General Form

The general form, $Ax + By = C$, of an equation is useful for writing linear equations without fractions. It is customary to write equations in general form with a positive coefficient on x.

Write each equation in general form.

35. $\frac{x}{3} + \frac{y}{5} = 1$

36. $y = \frac{1}{3}x + \frac{4}{3}$

37. The given table displays the dollar cost of a collect phone call based on the length of the call in minutes.

Minutes x	Cost y
3	$2.30
5	$3.00
10	$4.75
13	$5.80

(a) Determine the linear equation $f(x) = mx + b$ for the line that contains these data points.

(b) Determine the meaning of m and b in this application.

38. The given graph displays the dollar cost of having a clothes dryer repaired by a service shop.

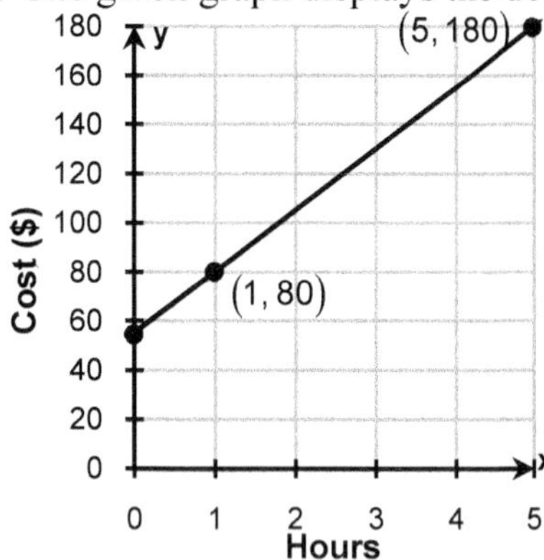

(a) Use Determine the linear equation $f(x) = mx + b$ for this line.

(b) Determine the meaning of m and b in this application.

3.3 Lecture Guide: Solving Systems of Linear Equations in Two Variables Graphically and Numerically

Objective 1: Check possible solutions of a linear system.

1. Determine whether the ordered pair $(2,-2)$ is a solution of the system of linear equations.

$$\begin{Bmatrix} y=-3x+2 \\ y=x-4 \end{Bmatrix}$$

2. Determine whether the ordered pair $(-1,-2)$ is a solution of the system of linear equations.

$$\begin{Bmatrix} 2x+y=-4 \\ -3x+y=1 \end{Bmatrix}$$

Objective 2: Solve a system of linear equations by using graphs and tables.

3. Consider the graph of the system: $\begin{Bmatrix} y=-\dfrac{x}{4}+2 \\ y=x-3 \end{Bmatrix}$

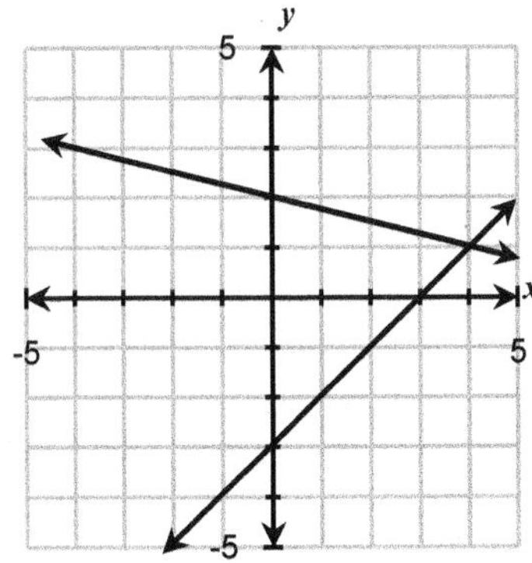

(a) Determine the point of intersection.

(b) Verify that this point satisfies both equations.

4. Use the table to solve the system of equations $\begin{Bmatrix} y_1 = 2x + 1 \\ y_2 = x - 5 \end{Bmatrix}$.

x	y_1	y_2
−9	−17	−14
−8	−15	−13
−7	−13	−12
−6	−11	−11
−5	−9	−10
−4	−7	−9
−3	−5	−8

Solution: ________________

5. Solve the system of equations $\begin{Bmatrix} x + y = 5 \\ x - 3y = -3 \end{Bmatrix}$ by graphing each line.

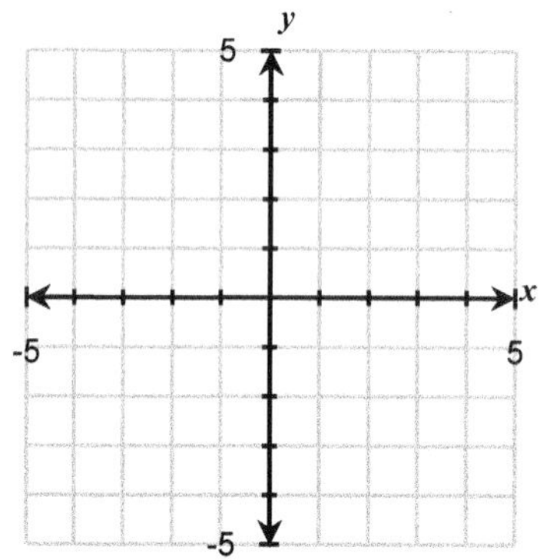

Solution: ________________

6. Solve the system of linear equations by using a calculator to graph each line and express both coordinates in fraction form. See Technology Perspective 3.3.1 for help.

$\begin{Bmatrix} y = 3x - 5 \\ y = -4x + 8 \end{Bmatrix}$

Solution: ________________

Objective 3: Identify inconsistent systems and systems of dependent linear equations.
Recall that we have three possibilities when solving linear equations in one variable: a ____________________ equation, an ____________________ or a ____________________.

7. Solve each equation and identify its type from the list above.

(a) $x+2=x+3$ **(b)** $x+2=x+2$ **(c)** $2x-1=x+1$

8. Sketch the graphs of the lines in each system below.

(a) $\begin{Bmatrix} y=x+2 \\ y=x+3 \end{Bmatrix}$ **(b)** 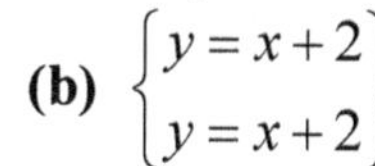 $\begin{Bmatrix} y=x+2 \\ y=x+2 \end{Bmatrix}$ **(c)** 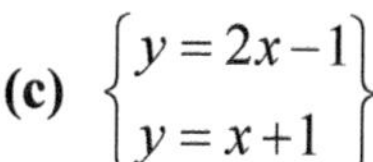 $\begin{Bmatrix} y=2x-1 \\ y=x+1 \end{Bmatrix}$

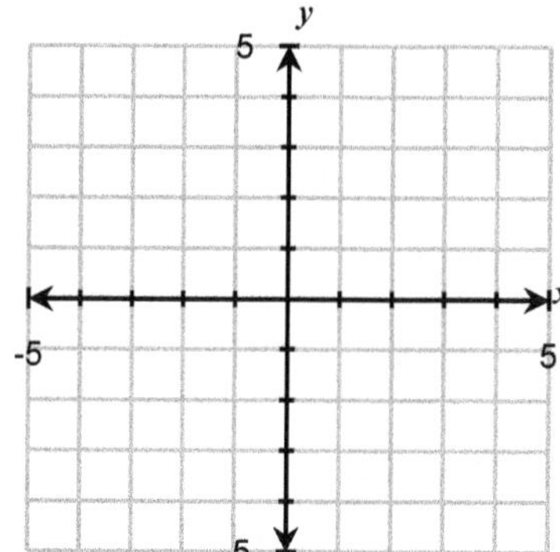

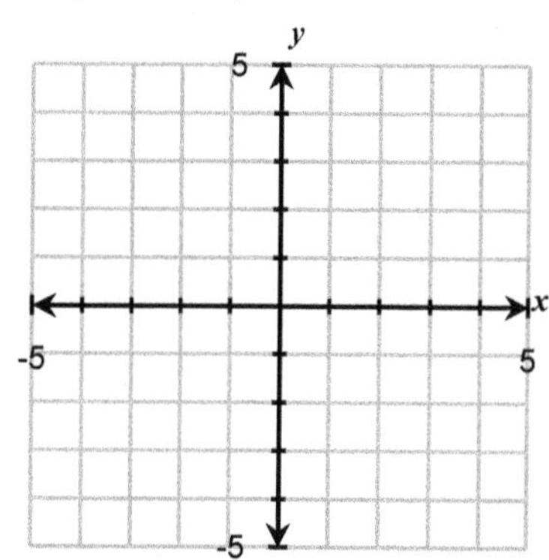

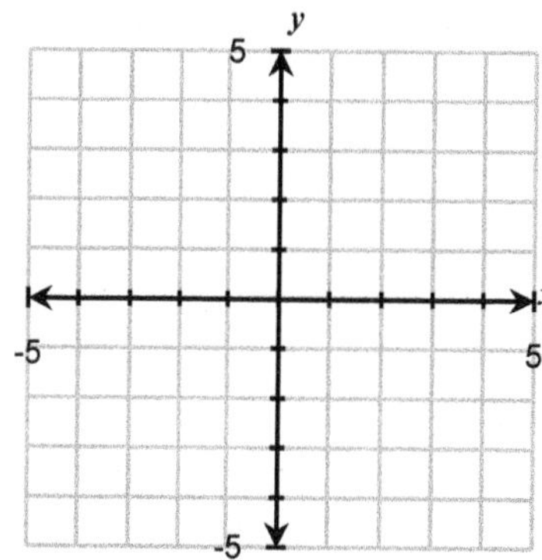

9. A solution of a system of equations is an ____________________ ____________________, (x, y), that satisfies each equation in the system.

10. Classify each system as one of the following: (Hint: Examine the graphs shown for problem 9.)

A. A consistent system of independent linear equations having exactly one solution.
B. An inconsistent system of linear equations having no solution.
C. A consistent system of dependent linear equations having an infinite number of solutions.

(a) $\begin{Bmatrix} y=x+2 \\ y=x+3 \end{Bmatrix}$ ________ **(b)** $\begin{Bmatrix} y=x+2 \\ y=x+2 \end{Bmatrix}$ ________ **(c)** $\begin{Bmatrix} y=2x-1 \\ y=x+1 \end{Bmatrix}$ ________

Use the slope-intercept form of each line to determine the number of solutions of the system and then classify each system as a consistent system of independent equations, an inconsistent system, or a consistent system of dependent equations. Then solve each system using a graph or a table.

11. $\begin{Bmatrix} y = \frac{2}{3}x + 4 \\ y = \frac{2}{3}x - 4 \end{Bmatrix}$

Consistent or inconsistent system?

Dependent or independent equations?

Solution: ____________________

12. $\begin{Bmatrix} y = -\frac{3}{4}x + 7 \\ y = \frac{3}{4}x - 2 \end{Bmatrix}$

Consistent or inconsistent system?

Dependent or independent equations?

Solution: ____________________

13. $\begin{Bmatrix} 5x - 2y = -10 \\ y = \frac{5}{2}x + 5 \end{Bmatrix}$

Consistent or inconsistent system?

Dependent or independent equations?

Solution: ____________________

14. $\begin{Bmatrix} x - y = 8 \\ x + y = 4 \end{Bmatrix}$

Consistent or inconsistent system?

Dependent or independent equations?

Solution: ____________________

15. The tables below display the charges for two taxi services based upon the number of miles driven. Service A has an initial charge of $2.30 and $0.15 for each quarter mile, while Service B has an initial charge of $2.00 and $0.20 for each quarter mile.

Service A

x Miles	y $ Cost
0.50	2.60
1.00	2.90
1.50	3.20
2.00	3.50
2.50	3.80
3.00	4.10

Service B

x Miles	y $ Cost
0.50	2.40
1.00	2.80
1.50	3.20
2.00	3.60
2.50	4.00
3.00	4.40

(a) Give the solution of the corresponding system of equations.

(b) Interpret the meaning of the x- and y-coordinates of this solution.

16. At the beginning of the semester a student receives a flier for two different phone plans. The first company offers a rate of 8 cents per minute with no additional fees. The second company offers a rate of 4 cents per minute but has a $20 per month fee in addition to any minutes used. In the past, this student has not used more than 800 minutes in a month. Write an equation to represent the cost y of each plan for one month in which x minutes were used.

Company A: $y_1 =$ ________________

Company B: $y_2 =$ ________________

(a) Determine the restrictions on the variable x.

(b) Use a calculator to graph the two equations. Use the **Zoom Fit** option to help you determine an appropriate viewing window. Draw a rough sketch of your calculator graph below. See Technology Perspective 3.3.2 for help.

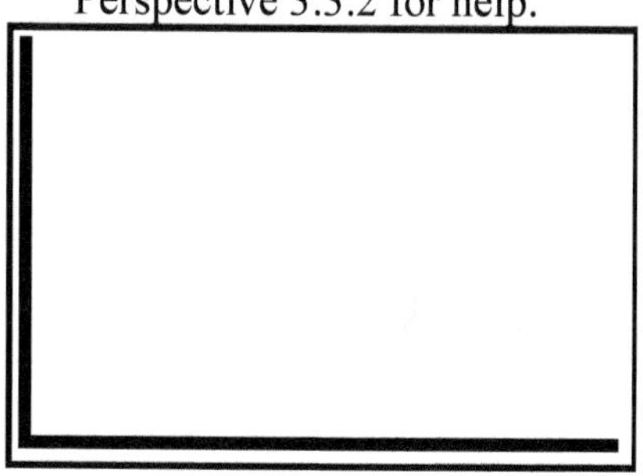

(c) Use a calculator to find the solution of the system.

(d) Interpret the meaning of the x- and y-coordinates of this solution.

3.4 Lecture Guide: Solving Systems of Linear Equations in Two Variables by the Substitution Method

Objective 1: Solve a system of linear equations by the substitution method.

Substitution Method

Step 1. Solve one of the equations for one ____________________ in terms of the other ____________________	Example: $\begin{cases} 6x+5y=12 \\ 2x+y=4 \end{cases}$
Step 2. ____________________ the expression obtained in Step 1 into the other equation (eliminating one of the variables), and solve the resulting equation.	
Step 3. Substitute the ____________________ obtained in Step 2 into the equation obtained in Step 1 (back-substitution) to find the value of the other variable.	
The ordered pair obtained in Steps 2 and 3 is the solution that should check in both equations.	

Solve each system using the substitution method.

1. $x=1+y$
$2x+y=5$

2. $3x+y=5$
$x-2y=4$

Solve each system using the substitution method.

3. $3x - 2y = -6$

$x - 4 = 0$

4. $\frac{x}{3} + \frac{y}{4} = 1$

$2y + \frac{1}{3}x = -6$

5. $y = -2x + 3$

$6x + 3y = 3$

6. $6x - 2y = 8$

$y = 3x - 4$

7. To review the three possible cases for systems of two linear equations in two variables, match each sentence with the case it describes.

(a) Consistent system of independent equations. ______

(b) Inconsistent system. ______

(c) Consistent system of dependent equations. ______

A. The solution process will produce unique *x*- and *y*-values.

B. The solution process will produce an identity.

C. The solution process will produce a contradiction.

8. The costs for renting a rug-shampooing machine from two different rental companies are given by the graphs shown below. The graph of $y_1 = f_1(x)$ gives the cost by Dependable Rental Company based upon the number of hours of use. The graph of $y_2 = f_2(x)$ gives the cost by Anytime Rental Company based upon the number of hours of use.

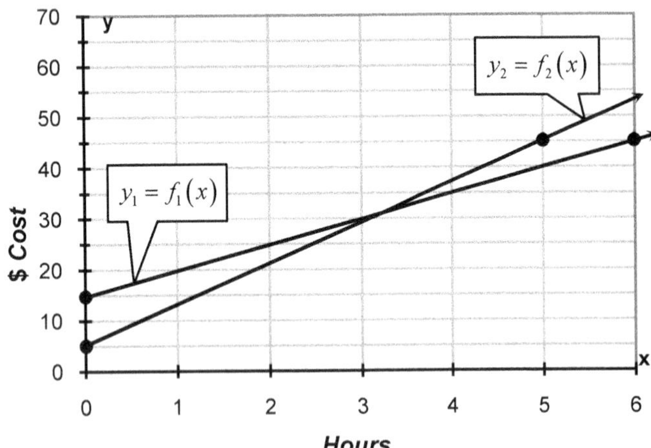

(a) Use the y-intercept and an additional point to determine the equation of the line for the Dependable Rental Company.

(b) Use the y-intercept and an additional point to determine the equation of the line for the Anytime Rental Company.

(c) Solve the system of equations using the substitution method.

(d) Interpret the meaning of the x- and y-coordinates of this solution.

3.5 Lecture Guide:
Solving Systems of Linear Equations in Two Variables by the Addition Method

Objective 1: Solve a system of linear equations by the addition method.

Addition Method

Step 1. Write both equations in the ________________ form $Ax + By = C$.

Example: $\begin{cases} 6x + 5y = 12 \\ 2x + y = 4 \end{cases}$

Step 2. If necessary, multiply each equation by a constant so that the equations have one variable for which the coefficients are additive ________________.

Step 3. Add the new equations to ________________ a variable, and then solve the resulting equivalent equation.

Step 4. ________________ this value into one of the original equations, and solve for the other variable.

The ordered pair obtained in Steps 3 and 4 is the solution that should check in both equations.

Solve each system using the addition method.

1. $x - y = 3$
$x + y = 5$

2. $2x - 7y = 1$
$x + 5y = 9$

Solve each system using the addition method.

3. $8x - 3y = 5$
$-5x + 7y = 43$

4. $\frac{x}{2} - \frac{y}{8} = 3$
$\frac{x}{4} + \frac{y}{2} = -3$

5. $4x - 8y = 20$
$3x - 6y = 15$

6. $5x - 6y = 4$
$10x - 12y = 5$

Solve each system using the either the substitution method or the addition method.

7. $3x - y = -14$
$5x + 2y = -5$

8. $2x + 3y = -2$
$4x - 5y = 7$

9. After the first two algebra exams, one student had a mean score of 90 and range of 12. What were the scores on the two exams?

(a) Identify the variables:

Let $x =$ the ____________ on one exam

Let $y =$ the ____________ on the other exam

(b) Write the word equations:

The ______________ of the two scores is 90.

The ______________ of the two scores is _____________.

(c) Translate the word equations into algebraic equations:

___________________ $= 90$

___________________ $=$ ____________

(d) Solve this system of equations:

(e) Write a sentence that answers the question:

(f) Is this answer reasonable?

3.6 Lecture Guide: More Applications of Linear Systems

Objective 1: Use systems of linear equations to solve word problems.

Strategy for Solving Word Problems

Step 1. Read the problem carefully to determine what you are being asked to find.
Step 2. Select a ___________ to represent each unknown quantity. Specify precisely what each variable represents and note any restrictions on each variable.
Step 3. If necessary, make a ___________ and translate the problem into a word equation or a system of word equations. Then translate each word equation into an ___________ equation.
Step 4. Solve the equation or system of equations, and answer the question completely in the form of a sentence.
Step 5. Check the ________________ of your answer.

1. The bill for a cellular phone in August was \$10 more than twice the September bill. The total that was required to pay both of these bills was \$175. What was the bill for each month?

(a) Identify the variables:
Let x = the cost of the phone bill for August

Let y = ____

(b) Write the word equations:
(1) August bill is ______ more than __________ the September bill

(2) The total of the _____________ bill and the _______________ bill is ___________ .

(c) Translate the word equations into algebraic equations:
(1) ________________________________

(2) ________________________________

(d) Solve this system of equations:

(e) Write a sentence that answers the question:

(f) Is this answer reasonable?

2. A hobbyist is making a picture frame out of wood molding. He has 42 inches of molding to create the frame. If the width of the frame must be 5 inches less than the length, what will be the dimensions of the frame? Will a 9 inch by 12 inch picture fit in this frame?

(a) Identify the variables (including the units of measurement):

Let L = the ____________________ of the picture frame in inches

Let W = the ____________________ of the picture frame in inches

(b) Write the word equations:

(1) The _________________ of the frame is __________ inches.

(2) The _______________ is 5 inches less than the __________________ .

(c) Translate the word equations into algebraic equations:

(1) 2L + ____________ = ____________

(2) W = ________________

(d) Solve this system of equations:

(e) Write a sentence that answers the question:

(f) Is this answer reasonable?

Rate Principle

Amount = *Rate* × *Base* $A = R \cdot B$

Applications of the Rate Principle

1. Variable cost = Cost per item × Number of items
2. Interest = Principal invested × Rate × Time
3. Distance = Rate × Time
4. Amount of active ingredient = Rate of concentration × Amount of mixture
5. Work = Rate × Time

Mixture Principle for Two Ingredients

Amount in first + Amount in second = Amount in mixture

Applications of the Mixture Principle

1. Amount of product A + Amount of product B = Total amount of mixture
2. Variable cost + Fixed cost = Total cost
3. Interest on bonds + Interest on CDs = Total interest
4. Distance by first plane + Distance by second plane = Total distance
5. Antifreeze in first solution + Antifreeze in second solution = Total amount of antifreeze

Solve each of the remaining problems using the word problem strategy illustrated in the first two problems.

3. A small t-shirt screening business operates on a daily fixed cost plus a variable cost that depends on the number of shirts screened in one day. The total cost for screening 260 shirts on Friday was \$1015. The total cost for screening 380 shirts on Saturday was \$1345. What is the fixed daily cost? What is the cost to screen each shirt?

4. The Candy Shop has two popular kinds of candy. The owner is trying to make a mixture of 100 pounds of these candies to sell at $3 per pound. If the gummy gums are priced at $2.50 per pound and the sweet treats are $3.75 per pound, how many pounds of each must be mixed in order to produce the desired amount?

5. Ashley invested money in two different accounts. One investment was at 10% simple interest and the other was at 12% simple interest. The amount invested at 10% was \$1500.00 more than the amount invested at 12%. The total interest earned was \$480.00. How much money did Ashley invest in each account?

6. A hospital needs 80 liters of a 12% solution of disinfectant. This solution is to be prepared from a 33% solution and a 5% solution. How many liters of each should be mixed to obtain this 12% solution?

7. Two brothers decided to get together one weekend for a visit. They live 472 miles apart. They both left their homes at 8:00 a.m. on Saturday and drove toward each other. The younger brother drove 6 mi/hr faster than the older brother and they met in 4 hours. How fast was each brother driving?

Lecture Guides for

Chapter 4

Linear Inequalities

4.1 Lecture Guide: Solving Linear Inequalities Using the Addition-Subtraction Principle

Objective 1: Identify linear inequalities and check a possible solution of an inequality.

Linear Inequalities

Verbally	**Algebraically**	**Algebraic Examples**	**Graphically**
A linear inequality in one variable is an inequality that is ____________ degree in that variable.	For real constants A, B, and C, with $A \neq 0$.		
	$Ax + B > C$	$x > 2$	(at 2, shaded right
	$Ax + B \geq C$	$x \geq 2$	[at 2, shaded right
	$Ax + B < C$	$x < 2$	) at 2, shaded left
	$Ax + B \leq C$	$x \leq 2$	] at 2, shaded left

1. Which of the following choices is a linear inequality in one variable?
   (a) $3x + 1 < 10$ **(b)** $3x + 1$ **(c)** $3x^2 + 1 > 10$ **(d)** $3x + 1 = 10$

Conditional Inequality

A **conditional inequality** contains a variable and is true for ____________, but not all, real values of the variable.

The **solution** of a linear inequality consists of all values that ____________ the inequality. The solution of a conditional linear inequality will be an interval that contains an infinite set of values.

2. Determine whether $x = 5$ satisfies each inequality.
   (a) $x < 5$ **(b)** $x \leq 5$ **(c)** $x > 5$ **(d)** $x \geq 5$

3. Determine whether either 4 or -4 satisfies the inequality $6x - 2 < 5x - 4$.

Objective 2: Solve linear inequalities in one variable using the addition-subtraction principle for inequalities.

Addition-Subtraction Principle for Inequalities

Verbally	**Algebraically**	**Numerical Example**
If the same number is ____________ to or subtracted from ____________ sides of an inequality, the result is an ________________ inequality.	If a, b, and c, are real numbers then a < b is equivalent to $a+c<b+c$ and to $a-c<b-c$	$x-2<5$ is equivalent to $x-2+2<5+2$ and to $x<7$.

Each inequality in problems 4-17 is a **conditional inequality.** Use the addition-subtraction principle of equality to solve each inequality. Give your answer in interval notation.

4. $x-4<11$

5. $3\le 3+x$

6. $3y\le 2y+7$

7. $7a\ge 6a-1$

8. $7d<6d$

9. $3x-1\le 2x+6$

10. $8x-2>7x+12$

11. $5-3x<6-4x$

12. $6(2x-4)>11x+8$

13. $7(2x-2)<13(x+1)$

14. $3(2m+1)\geq 5(m-1)-5$

15. $3(x+2)\geq 2(2x-5)$

16. $-6(2-x)\leq 4x-3(5-x)$

17. $\frac{y}{4}-\frac{3}{5}>\frac{7}{5}+\frac{5y}{4}$

Objective 3: Use graphs and tables to solve linear inequalities in one variable.

18. The graph below displays the monthly cost y of two phone plans based on x minutes of use. The graph of y_1 represents the monthly cost of plan A and the graph of y_2 represents the monthly cost of plan B.

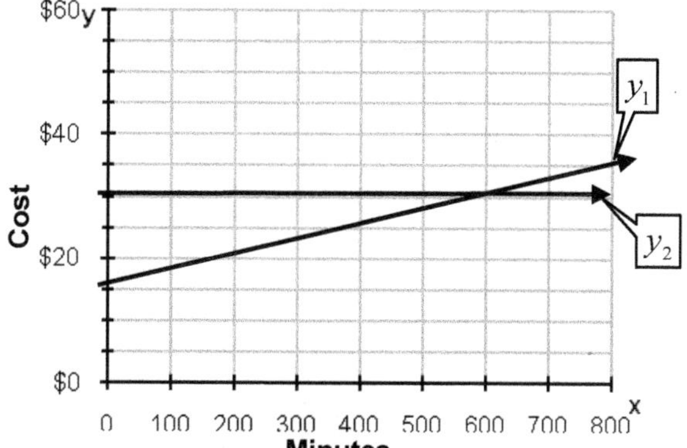

(a) Approximate the monthly cost of plan A with 400 minutes of use.

(b) Approximate the monthly cost of plan B with 400 minutes of use.

(c) Approximate the monthly cost of plan A with 800 minutes of use.

(d) Approximate the monthly cost of plan B with 800 minutes of use.

(e) For how many minutes of use for will both plans have the same monthly cost?

(f) What is that monthly cost?

(g) Explain the circumstances under which you would choose plan A.

(h) Explain the circumstances under which you would choose plan B.

19. Use the graph to solve each equation or inequality.

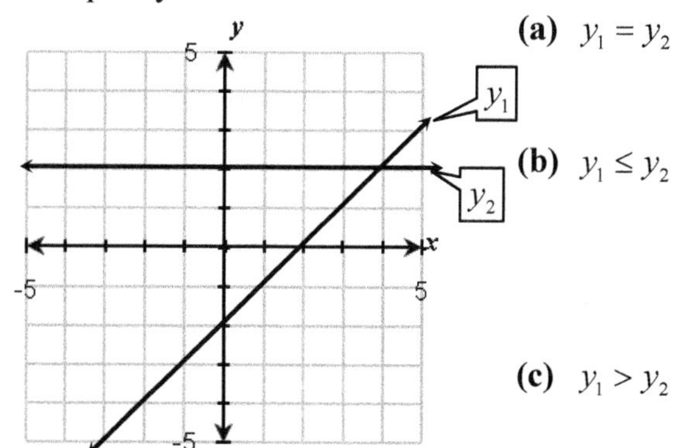

(a) $y_1 = y_2$

(b) $y_1 \le y_2$

(c) $y_1 > y_2$

20. Use the graph to solve each equation or inequality.

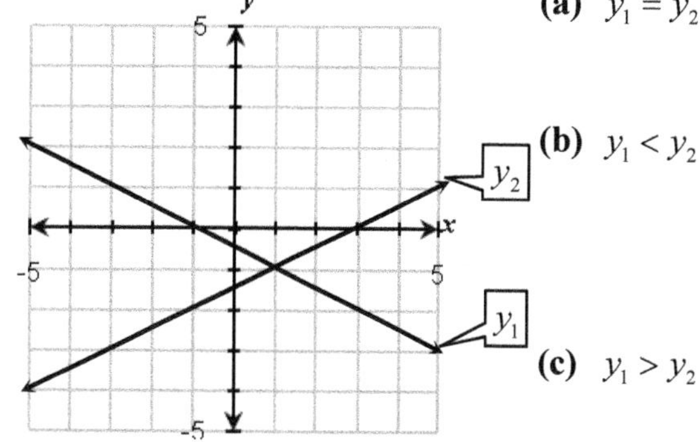

(a) $y_1 = y_2$

(b) $y_1 < y_2$

(c) $y_1 > y_2$

21. Use the table to solve each equation or inequality.

x	y_1	<, =, or >	y_2
–4	1		–2
–3	2		0
–2	3		2
–1	4		4
0	5		6
1	6		8
2	7		10

(a) $y_1 = y_2$

(b) $y_1 < y_2$

(c) $y_1 \geq y_2$

22. Use the table to solve each equation or inequality.

x	y_1	<, =, or >	y_2
4	–2		2
5	1		4
6	4		6
7	7		8
8	10		10
9	13		12
10	16		14

(a) $y_1 = y_2$

(b) $y_1 \leq y_2$

(c) $y_1 \geq y_2$

23. Solve the inequality $4x - 5 > 3x - 2$ by letting $Y_1 = 4x - 5$ and $Y_2 = 3x - 2$.

(a) Use your calculator to create a graph of Y_1 and Y_2 using a viewing window of $[-2, 6, 1]$ by $[-5, 10, 1]$. Use the **Intersect** feature to find the point where these two lines intersect. Draw a rough sketch below. The values in the table will help.

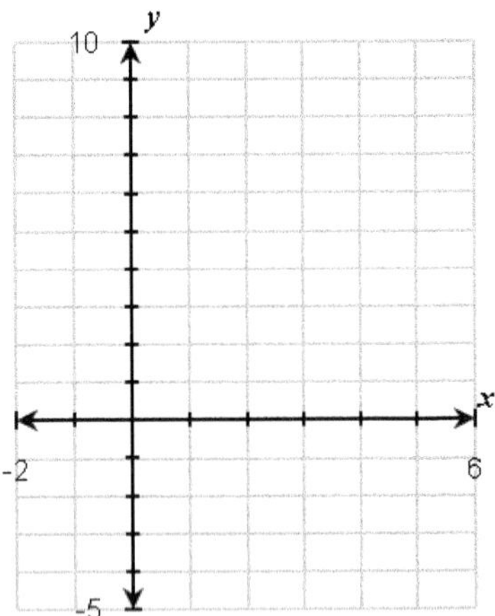

(b) Y_1 is above Y_2 for x-values to the ____________ of ______.

(c) Create a table on your calculator with the table settings: TblStart = 0; ΔTbl = 1
Complete the table below.

x	Y_1	<, =, or >	Y_2
0			
1			
2			
3			
4			
5			
6			

$Y_1 > Y_2$ for x-values ____________ than ______.

(d) Solve the inequality $4x - 5 > 3x - 2$ algebraically.

(e) Do your solutions all match?

Solve each inequality algebraically or graphically.

24. $0.5x + 3 > -0.5x - 5$

25. $2(x+3) - 3x \le 5 - 2x$

26. The tables below display the charges for two taxi services based upon the number of miles driven. Service A has an initial charge of \$2.30 and \$0.15 for each quarter mile, while Service B has an initial charge of \$2.00 and \$0.20 for each quarter mile.

Service A

x Miles	y_1 \$ Cost
0.50	2.60
1.00	2.90
1.50	3.20
2.00	3.50
2.50	3.80
3.00	4.10

Service B

x Miles	y_2 \$ Cost
0.50	2.40
1.00	2.80
1.50	3.20
2.00	3.60
2.50	4.00
3.00	4.40

Use these tables to solve

(a) $y_1 = y_2$

(b) $y_1 < y_2$

(c) $y_1 > y_2$

(d) Interpret the meaning of each solution in parts **a – c**.

27. Complete the following table.

Phrase	**Inequality Notation**	**Interval Notation**	**Graphical Notation**
"x is at least 5"			⟷
"x is at most 2"			⟷
"x exceeds -3"			⟷
"x is smaller than -1"			⟷

28. Write an algebraic inequality for the following statement, using the variable x to represent the number, and then solve for x.

Verbal Statement:
Five less than three times a number is at least two times the sum of the number and three.

Algebraic Inequality:

Solve this inequality:

29. The perimeter of the triangle shown must be less than 26 cm.
Find the possible values for a.

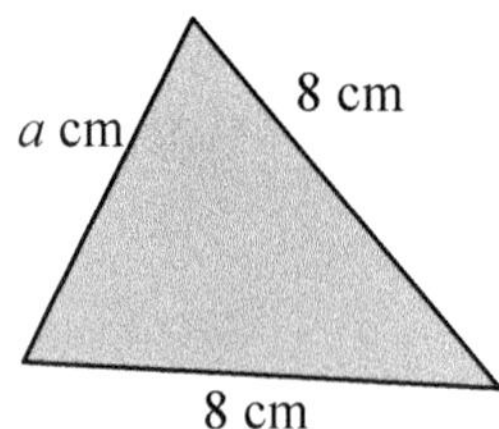

4.2 Lecture Guide: Solving Linear Inequalities Using the Multiplication-Division Principle

Objective 1: Solve linear inequalities using the multiplication-division principle for inequalities.

Multiplication-Division Principle for Inequalities:

Verbally	Algebraically	Numerical Examples
Order Preserving: If ___________ sides of an inequality are multiplied or ___________ by a *positive number*, the result is an inequality that has the same ___________ as the original inequality.	If *a*, *b*, and *c* are real numbers and $c > 0$, then $a > b$ is ___________ to $ac > bc$.	$\frac{x}{2} > 3$ is equivalent to $2\left(\frac{x}{2}\right) > 2(3)$ and to $x > 6$
Order Reversing: If ___________ sides of an inequality are multiplied or divided by a negative number and the order of inequality is ___________, the result is an inequality that has the ___________ solution as the original inequality.	If *a*, *b*, and *c* are real numbers and $c < 0$, then $a > b$ is equivalent to $ac < bc$.	$-\frac{x}{3} > 5$ is equivalent to $(-3)\left(-\frac{x}{3}\right) < (-3)(5)$ and to $x < -15$

Use the multiplication-division principle of equality to solve each inequality.

1. $8x < 72$

2. $-a > 12$

3. $\frac{x}{5} \geq -4$

4. $7x + 5 \leq -23$

5. $2(3x-1)>2x+22$

6. $3(4x-5)<5(2x+1)$

7. $8\geq\frac{-2}{5}x$

8. $\frac{x}{3}<\frac{x}{2}+5$

9. $-3(x-4)\leq 6-(x+2)$

10. $2(3x-5)-2x>3x-15(x-2)$

11. Use the table to solve each equation or inequality.

x	y_1	y_2
5	−9	11
6	−6	9
7	−3	7
8	0	5
9	3	3
10	6	1
11	9	−1

(a) $y_1 = y_2$

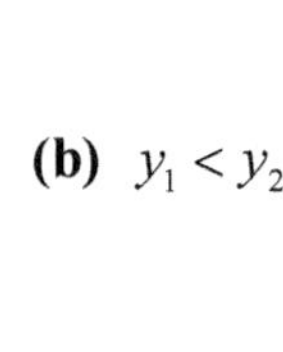

(b) $y_1 < y_2$

(c) $y_1 \geq y_2$

12. Use the graph to solve each equation or inequality.

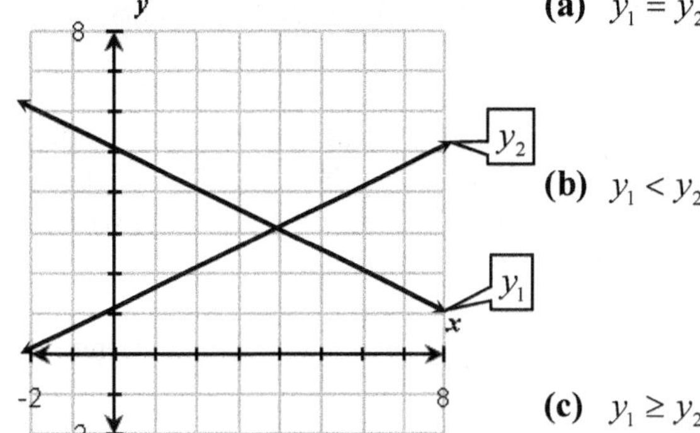

(a) $y_1 = y_2$

(b) $y_1 < y_2$

(c) $y_1 \geq y_2$

13. A high school band is having a fund raiser to raise money for a trip. The cost of renting a snow-cone machine for the fundraiser includes a fixed cost of \$84 plus a variable cost of \$0.30 per snow-cone. Snow-cones can be sold for \$1.50 each.

(a) Write an equation for the cost y_1 of renting the machine and selling x snow-cones.

(b) Write an equation for the revenue y_2 generated by selling x snow-cones.

(c) Determine the values of x for which $y_1 < y_2$.

(d) Interpret the meaning of the answer to part **(c)**.

4.3 Lecture Guide: Solving Compound Inequalities

Objective 1: Identify an inequality that is a contradiction or an unconditional inequality.

The algebraic process for solving the inequalities we have examined in the first two sections of this chapter has left a variable term on one side of the inequality. These have all been **conditional inequalities**. Sometimes the algebraic process for solving an inequality will result in the variable being completely removed from the inequality, which means the inequality is a **contradiction** or an **unconditional inequality**.

A ________________ inequality is an inequality that is true for some but not all values of the variable.

An ____________________ inequality is an inequality that is true for all values of the variable.

A ________________ is an inequality that is not true for any value of the variable.

1. Identify each inequality as a conditional inequality, an unconditional inequality, or a contradiction.
(a) $x+2<x$ **(b)** $x+2>x-3$ **(c)** $x+2<-3$

Solve each inequality. Identify each contradiction or unconditional inequality.

2. $3x+2-x\geq 4+2x$ **3.** $5x+3-2x\leq 3x+3$

Identify each inequality as a **conditional inequality**, a **contradiction** or an **unconditional inequality** and solve.

4. $5x-2-7x \geq 2(4-x)$

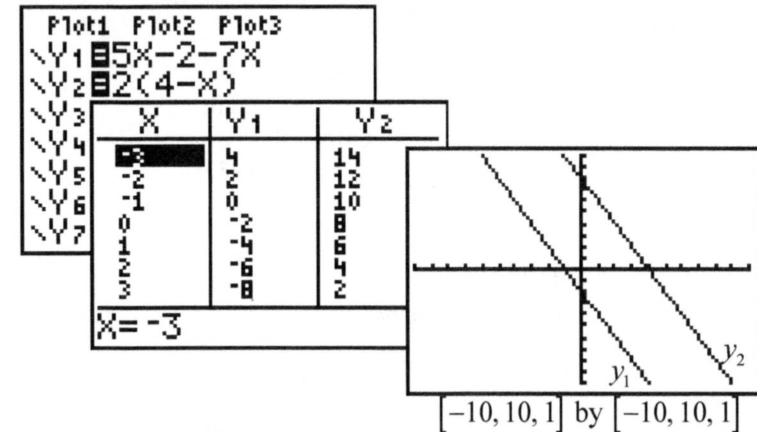

$[-10, 10, 1]$ by $[-10, 10, 1]$

Type:

Solution:

5. $2x-(x-4) > 5x-6(x+1)$

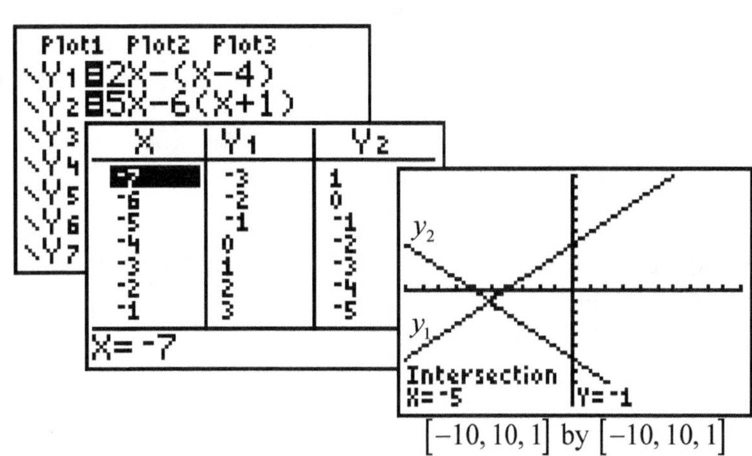

$[-10, 10, 1]$ by $[-10, 10, 1]$

Type:

Solution:

6. $3(6x-5) < 3x+15+15x$

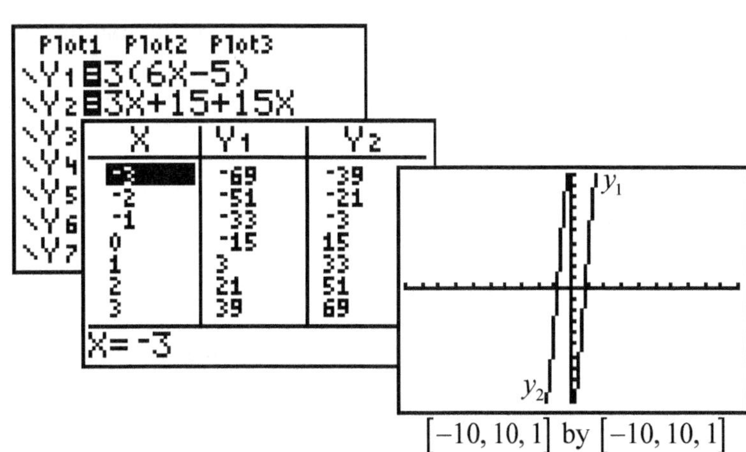

$[-10, 10, 1]$ by $[-10, 10, 1]$

Type:

Solution:

Objective 2: Write the intersection or union of two intervals.

Intersection of Two Sets

Algebraic Notation	Verbally	Numerical Example	Graphical Example
$A \cap B$	The intersection of A and B is the set that contains the elements in ____________ *A and B.*	$(-3,4) \cap [0,6]$ $= [0,4)$	

Union of Two Sets

Algebraic Notation	Verbally	Numerical Example	Graphical Example
$A \cup B$	The ____________ of A and B is the set that contains the elements in either *A or B or both.*	$(-3,4) \cup [0,6]$ $= (-3,6]$	

Write each inequality as two separate inequalities using the word "and" to connect the inequalities.

7. $0 \le x < 2$

8. $-13 < x \le -3$

Write each inequality expression as a single compound inequality.

9. $x > -2$ and $x < 3$

10. $x \ge -10$ and $x < -8$

Represent each union of intervals by two separate inequalities, using the word *or* to connect the inequalities.

11. $(-\infty,-4) \cup (3,\infty)$

12. $(-1,2) \cup [5,9]$

Consider the sets $A = \{0,2,4,5,6\}$ and $B = \{2,4,6,8,10\}$.

13. Determine $A \cap B$.

14. Determine $A \cup B$.

15. (a) Using the word ____________ between two inequalities indicates the intersection of two sets. In some cases, an intersection can be written in a combined form that looks like one expression "sandwiched" between two other expressions.

(b) Using the word ____________ between two inequalities indicates the union of two sets.

Graph each pair of intervals on the same number line and then give both their intersection $A \cap B$ and their union $A \cup B$.

16. $A=(3,6]$; $B=[-1,4]$

$A \cap B =$ ____________________

$A \cup B =$ ____________________

17. $A=(-3,9)$; $B=[2,17)$

$A \cap B =$ ____________________

$A \cup B =$ ____________________

18. $A=(-\infty,6]$; $B=[1,\infty)$

$A \cap B =$ ____________________

$A \cup B =$ ____________________

19. $A=(-\infty,2)$; $B=(5,\infty)$

$A \cap B =$ ____________________

$A \cup B =$ ____________________

20. Complete the following table.

Compound Inequality	Verbal Description	Graph	Interval Notation
$x>3$ and $x \le 7$			
$-1<x \le 5$			
$x<-1$ or $x>2$			

Objective 3: Solve compound inequalities involving intersection and union.

Solve each compound inequality. Give the solution in interval notation.

21. $-12 \le 6x < 24$

22. $-1 < 2x + 1 \le 3$

23. $-1 < \frac{x}{2} + 3 \le 4$

24. $5x - 2 \le 6x + 1 \le 5x + 4$

25. $3x + 2 \le x - 1$ or $4x - 1 > 3(x - 2)$

26. $3x-2\geq 6-x$ and $3x-2\leq 12-x$

27. $2x-3<5x+9$ or $\frac{x}{2}\leq\frac{x}{3}-1$

28. $5x-1\geq 3x+9$ and $4x+2>7x-16$

29. Use the graph below to determine the solution of $-\frac{x}{3}-5 \leq \frac{2}{3}x+1 \leq -\frac{x}{3}+6$.

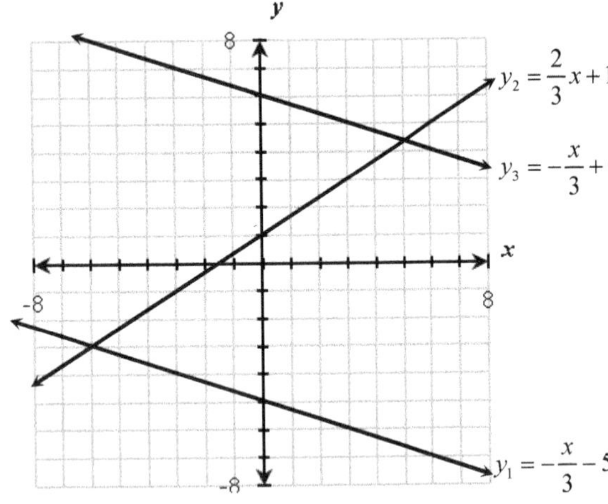

Solution: ____________________

30. A golfer must hit a shot more than 170 yd to clear a lake in front of a green. The depth of the green is only 20 yd. Write a compound inequality that expresses the length of the tee shot that a golfer must make to clear the water and to land safely on the green of this golf hole.

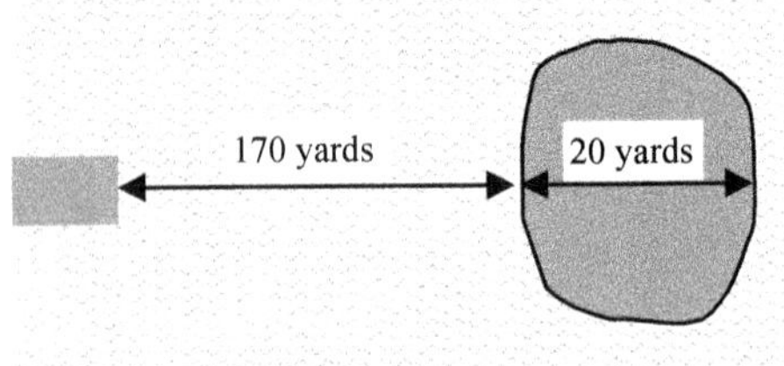

31. The perimeter of the parallelogram shown must be at least 20 cm and no more than 48 cm. If the given length must be 5 cm, determine the possible lengths for x, the unknown dimension.

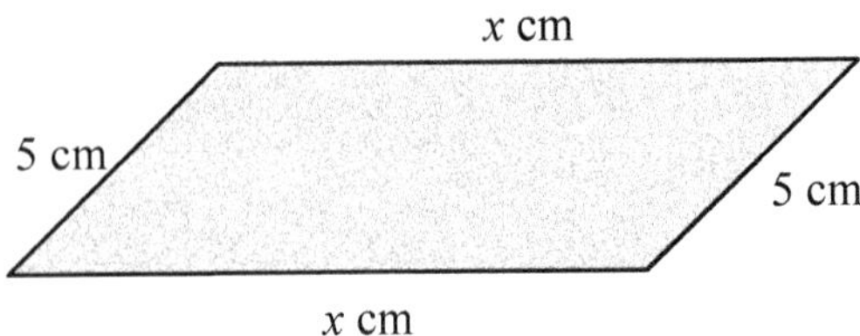

4.4 Lecture Guide: Solving Absolute Value Equations and Inequalities

Objective 1: Use absolute value notation to represent intervals.

Absolute Value Expressions

For any real number x and any nonnegative real number d:

Algebraically	Verbally	Algebraic Example	Graphical Example
$\|x\| = d$	The distance from 0 to x is d units.	$\|x\| = 3$ if $x = -3$ or $x = 3$	−3 0 3
$\|x\| < d$	The distance from 0 to x is ______________ than d units.	$\|x\| < 3$ if $-3 < x < 3$ $x > -3$ and $x < 3$	−3 0 3
$\|x\| > d$	The distance from 0 to x is ______________ than d units.	$\|x\| > 3$ if $x < -3$ or $x > 3$	−3 0 3

Use interval notation to represent the real numbers that are solutions of these inequalities.

1. $|x| < 9$

2. $|x| \geq 9$

Write an absolute-value inequality to represent each set of points.

3.

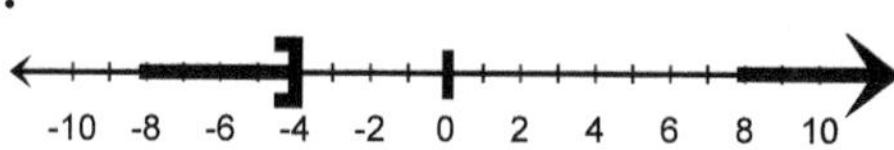

4.

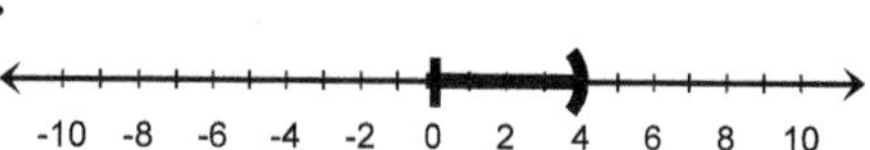

5. The points between −6 and 6.

6. $(-\infty, -1) \cup (1, \infty)$

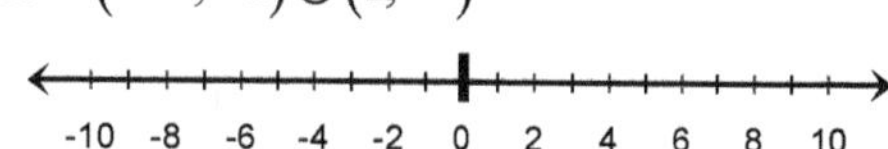

Use absolute value to represent the distance between these real numbers.

7. y and x

8. x and 10

9. x and -10

10. Write an absolute value equation indicating that the distance from x to a is d units.

Objective 2: Solve absolute value equations and inequalities.

Solving Absolute Value Equations and Inequalities

For any real numbers x and a and positive real number d:

Absolute Value Expression	**Verbally**	**Graphically**	**Equivalent Expression**
$\|x-a\|=d$	x is either d units ______________ or ______________ of a.	$a-d$ a $a+d$	$x-a=-d$ or $x-a=+d$
$\|x-a\|<d$	x is ______________ than d units from a.	$a-d$ a $a+d$	$-d<x-a<+d$
$\|x-a\|>d$	x is ______________ than d units from a.	$a-d$ a $a+d$	$x-a>d$ or $x-a<-d$

Similar statements can also be made about the order relations less than or equal to $(\le)$ and greater than or equal to $(\ge)$. Expressions with d negative are examined in the group exercises at the end of this section.

Write an absolute-value inequality to represent each interval. First graph the interval and use this graph to assist you in writing the inequality.

11. $(-\infty,-2]\cup[8,\infty)$

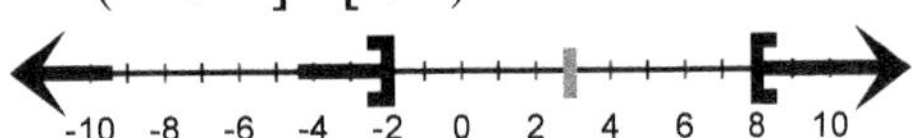

12. $(-5,3)$

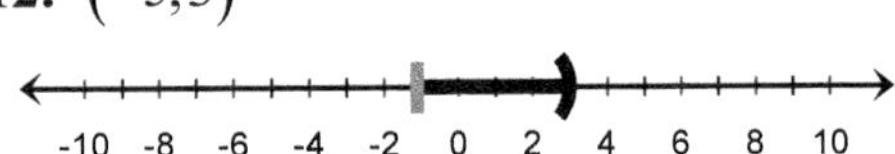

Solve each equation or inequality algebraically.

13. $|x-3|=8$

14. $|x+3|\le 10$

15. $|2x| < 22$

16. $|2x - 1| > 9$

17. $\left|\frac{2x}{3}\right| \le 12$

18. $|2(x+1) - 5| \ge 13$

19. $|x + 5| + 1 > 9$

20. $|3x - 5| - 2 = 8$

If $|a| = |b|$, then a and b are equal in magnitude but their signs can either agree or disagree. Thus $|a| = |b|$ is equivalent to ______________ or ______________. Use this result to solve the next two problems.

21. Solve $|x - 1| = |2x|$.

22. Solve $|3x - 1| = |2x + 3|$.

23. Use the graph to solve each equation or inequality.

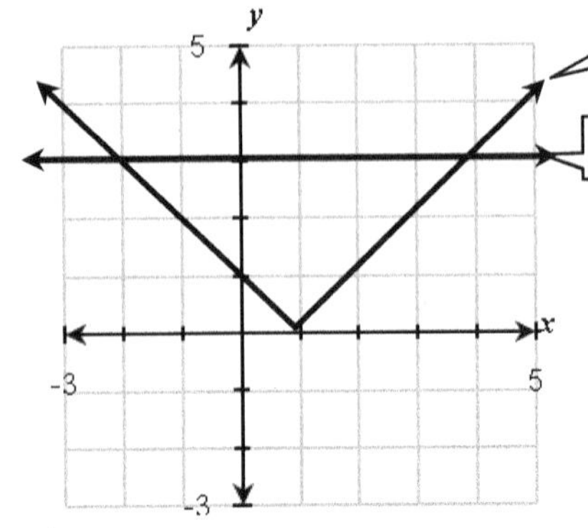

(a) $y_1 = y_2$

(b) $y_1 \leq y_2$

(c) $y_1 > y_2$

24. Use the table to solve each equation or inequality.

x	y_1	<, =, or >	y_2
−8	3		2
−7	2		2
−6	1		2
−5	0		2
−4	1		2
−3	2		2
−2	3		2

(a) $y_1 = y_2$

(b) $y_1 \le y_2$

(c) $y_1 > y_2$

Solve each equation or inequality algebraically. Use the given graph to solve each equation or inequality graphically. Then complete the table to check your solutions numerically.

25. $|x-2| = 5$

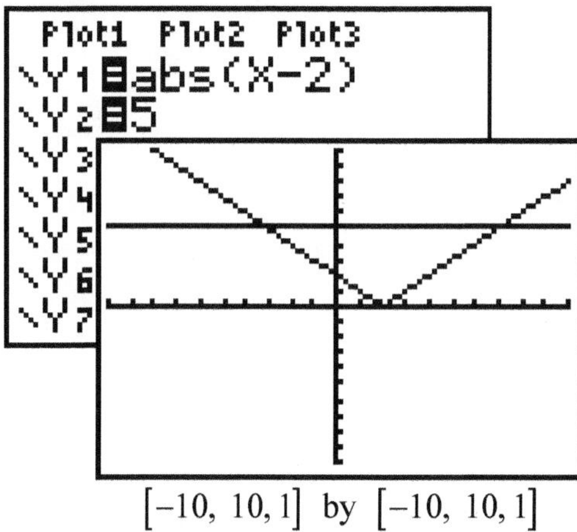

$[-10, 10, 1]$ by $[-10, 10, 1]$

26. $|x-2| \le 5$

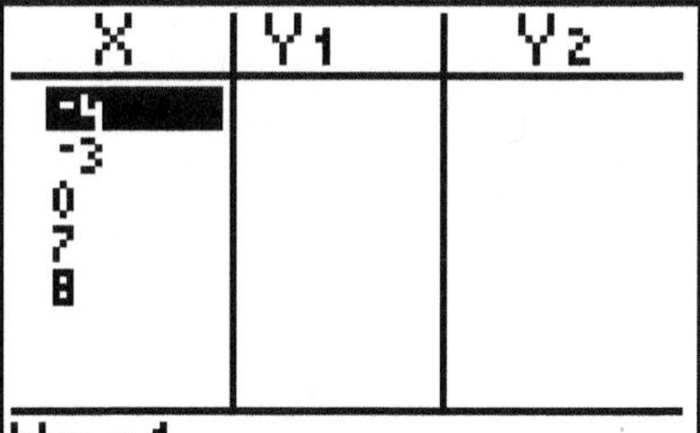

X	Y1	Y2
-4		
-3		
0		
7		
8		

X= -4

27. $|x-2| \ge 5$

Solve each equation or inequality algebraically. Use the given graph to solve each equation or inequality graphically. Then complete the table to check your solutions numerically.

28. $|2x-3|=5$

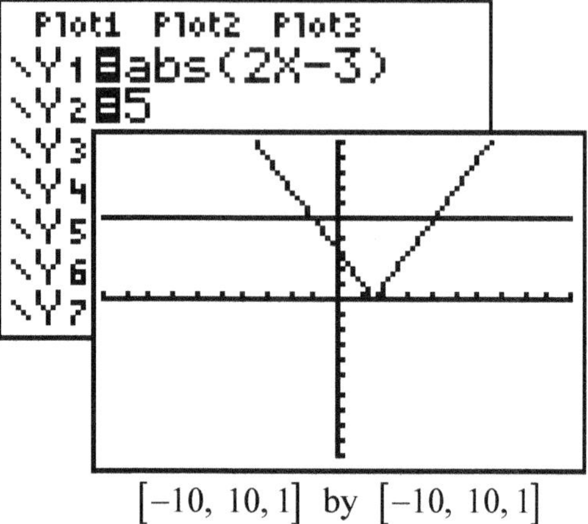

$[-10, 10, 1]$ by $[-10, 10, 1]$

X	Y1	Y2
-2		
-1		
0		
4		
5		

X= -2

29. $|2x-3|\geq 5$

30. $|2x-3|<5$

Write an absolute value inequality to represent the following intervals. Hint: First sketch a graph of these inequalities.

31. $(-1,15)$

32. $(-\infty,-16]\cup[6,\infty)$

33. $(-\infty,-9)\cup(-1,\infty)$

34. $[21,45]$

35. The correct torque setting for the lug bolts on a race car is 85 foot-pounds with a tolerance of ± 3 foot-pounds.

(a) Express the acceptable torque setting as an absolute value inequality.

(b) Express the acceptable torque setting as a compound linear inequality.

(c) Determine the lower and upper limits of the interval.

4.5 Lecture Guide: Graphing Systems of Linear Inequalities in Two Variables

Objective 1: Identify a solution of a linear inequality in two variables.

A **solution of a linear inequality** in two variables is an ordered pair of values that, when substituted into the inequality, makes a true statement.

1. Determine whether each point is a solution of $5x + y > 10$.

(a) $(2,1)$ **(b)** $(2,0)$ **(c)** $(0,0)$

Objective 2: Graph a linear inequality in two variables.

Graphing a Linear Inequality

Step 1. Graph the equality $Ax + By = C$ using

(a) A ________________ line for $\leq$ or $\geq$.

(b) A ________________ line for $<$ or $>$.

Step 2. Choose an arbitrary ______________ point not on the line; $(0,0)$ is often convenient. Substitute this test point into the inequality.

Step 3. **(a)** If the test point ______________ the inequality, shade the half-plane containing this point.

(b) If the test point does not satisfy the inequality, shade the ______________ half-plane.

Graph each linear inequality. Where possible, use your calculator as a check.
See Technology Perspective 4.5.1.

2. $y < \frac{2}{5}x - 3$

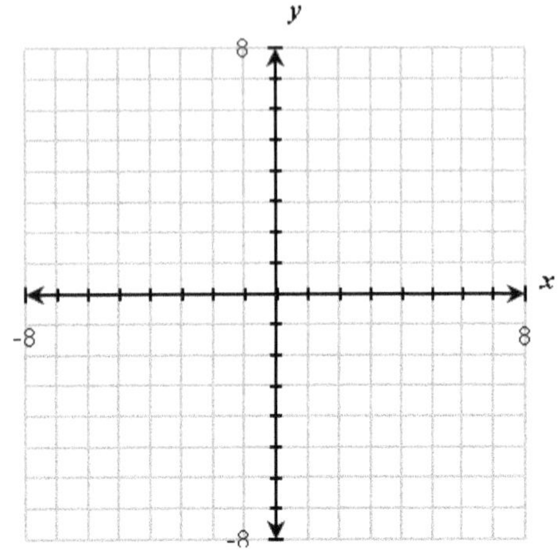

3. $3x + 4y \leq 16$

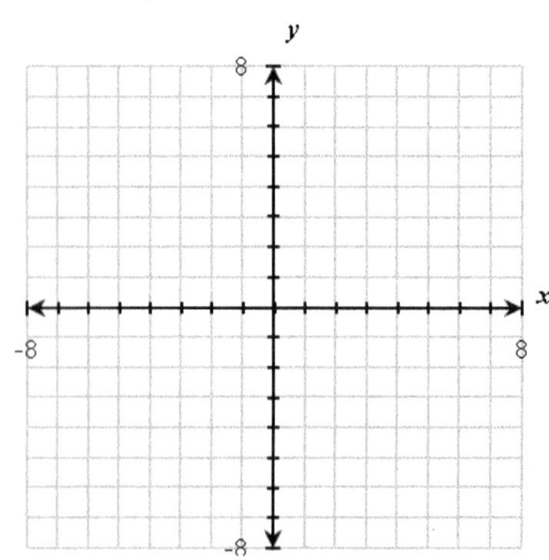

4. $\frac{x}{4}+\frac{y}{3}\leq 1$

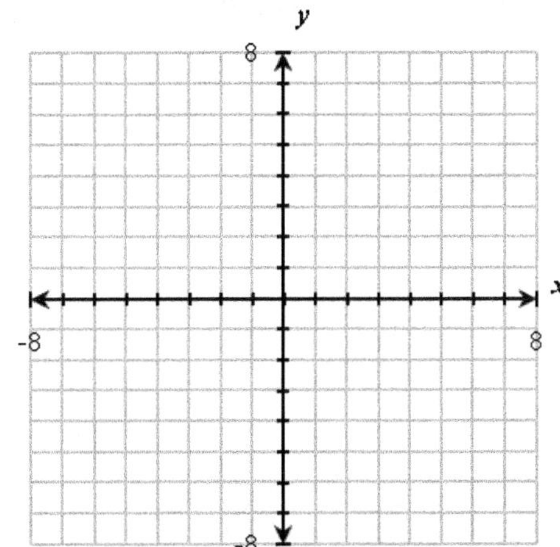

5. $3x-2y>6$

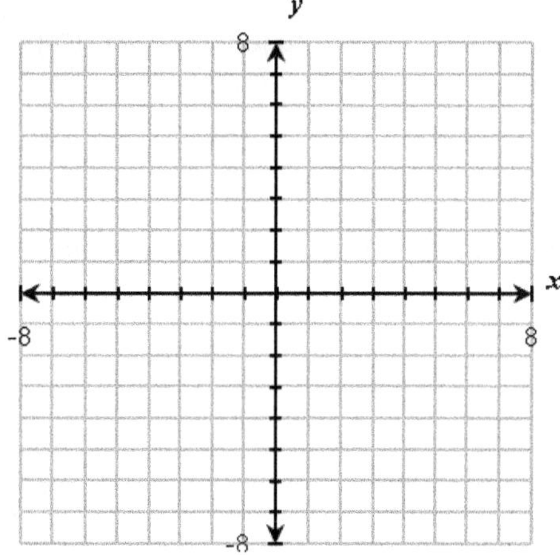

6. $y<-3$

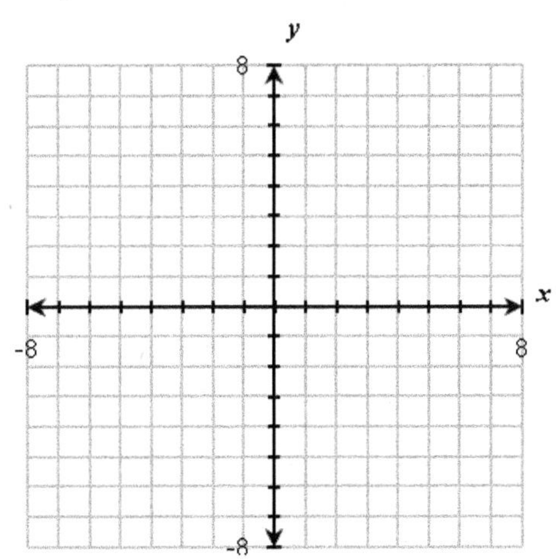

7. $x\geq -2$

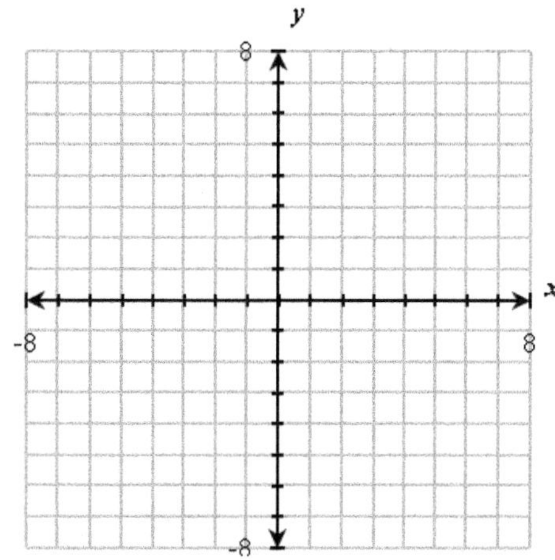

Objective 3: Graph a system of linear inequalities.

To graph a system of linear inequalities, we graph each inequality. Then the solution of the system of inequalities is the intersection of the regions shaded from each inequality. This is usually represented by shading this region.

Graph the solution of each system of linear inequalities.

8. $y - x > 1$
$y + 2x \le 4$

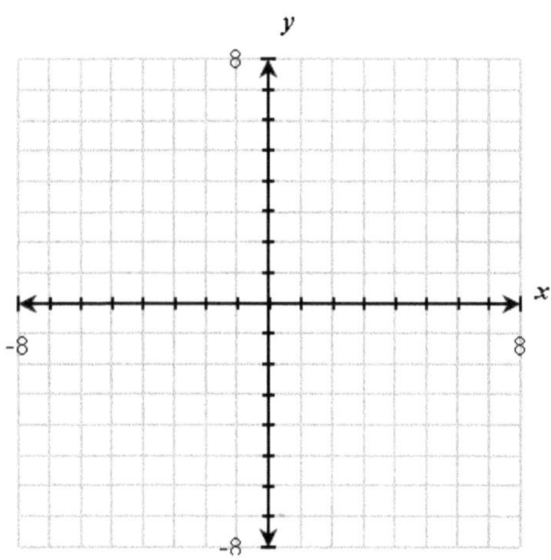

9. $2x - y \le 6$
$x + 2y < 10$

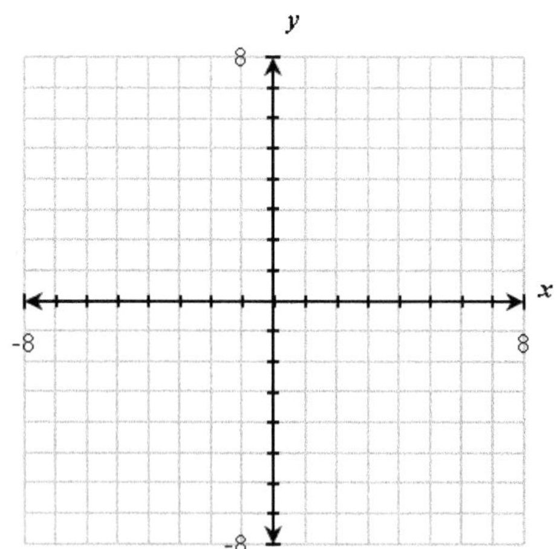

10.
$$x \geq 0$$
$$y \geq 0$$
$$2x + y - 8 \leq 0$$
$$-x + y \leq 2$$

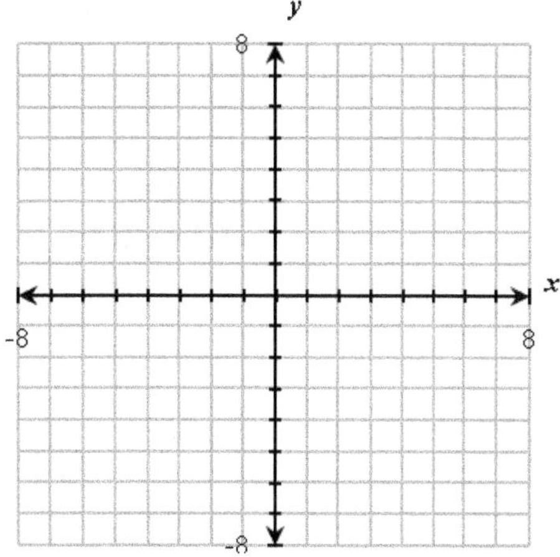

11. A local school sells student and adult tickets for basketball games. The number of student tickets x and adult tickets y sold each game are both nonnegative, and the gym holds a maximum of 2000 people. Write a system of inequalities that represents this situation and graph the solution of this system.

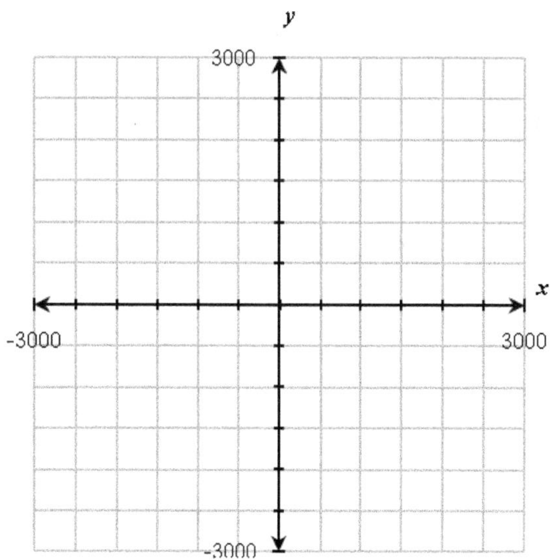

Lecture Guides for

Chapter 5

Exponents and Operations with Polynomials

5.1 Lecture Guide: Product and Power Rules for Exponents

Objective 1: Convert between exponential form and expanded form.

Exponential Notation

Algebraically	**Verbally**	**Numerical Examples**
For any natural number n, $b^n = \underbrace{b \cdot b \cdot \ldots \cdot b}_{n \text{ factors of } b}$ with base b and exponent n.	For any natural number n, b^n is the product of b used as a ____________ n times. The expression b^n is read as "b to the nth power."	$5^3 = 5 \cdot 5 \cdot 5$ $(-4)^2 = (-4)(-4)$

Write each expression in exponential form.

1. $x \cdot x \cdot x \cdot y \cdot y$

2. $-2 \cdot x \cdot x \cdot x \cdot x$

3. $x \cdot x \cdot x + y \cdot y \cdot y$

4. $(x+y)(x+y)(x+y)$

Write each exponential expression in expanded form.

5. $-5a^3$

6. $(-5a)^3$

7. a^3b^2

8. -7^2

9. $a^3 + b^2$

10. $(a+b)^2$

Objective 2: Use the product rule for exponents.

11. Complete the warm-up examples below:

Expanded Form:

$x^2 \cdot x^5 = (x \cdot x)(x \cdot x \cdot x \cdot x \cdot x)$

$=$ ____________________

$=$ ______

Alternate form:

$x^2 \cdot x^5 = x^{2+5}$

$=$ ______

Product Rule for Exponents

Algebraically	**Verbally**	**Algebraic Example**
For any real number x and natural numbers m and n, $x^m \cdot x^n = x^{m+n}$	To multiply two factors with the same base, use the common base and ____________ the exponents.	$x^2 \cdot x^4 =$ ____________

Simplify each expression

12. m^4m^3

13. $b^{10}b^{12}$

14. $(2x^3)(3x^5)$

15. $-r^4(r^6)(3r^2)$

Objective 3: Use the power rule for exponents.

16. Complete the warm-up examples below:

Expanded Form:

$(x^2)^3 = (x^2)(x^2)(x^2)$

$=$ ____________________

$=$ ______

Alternate form:

$(x^2)^3 = x^{2\cdot 3}$

$=$ ______

Power Rule For Exponents

Algebraically	**Verbally**	**Algebraic Example**
For any real number x, and natural numbers m and n, $(x^m)^n = x^{mn}$	To raise a power to a power, ____________ the exponents.	$(x^2)^5 =$ ____________

Simplify each expression.

17. $(x^3)^4$

18. $(-x^2)^3$

19. $(-x^3)^2$

20. $-(x^3)^2$

21. Complete the warm-up examples below:

Expanded Form:

$(xy)^3 = (xy)(xy)(xy)$

$= xxx \cdot yyy$

$=$ ________

Alternate form:

$(xy)^3 =$ ________

Raising Products and Quotients to a Power

Algebraically	Verbally	Algebraic Example
For any real numbers x and y and any natural number m, $(xy)^m = x^m y^m$	To raise a product to a power, raise each ________ to this power.	$(xy)^5 =$ ________
$\left(\frac{x}{y}\right)^m = \frac{x^m}{y^m}$ for $y \neq 0$	To raise a quotient to a power, raise both the ________ and the ________ to this power.	$\left(\frac{x}{y}\right)^5 =$ ________

Simplify each expression. Assume that variables are restricted to values that prevent division by zero.

22. $(mn)^3$

23. $(x^2y)^4$

24. $\left(\frac{x}{y}\right)^3$

25. $\left(\frac{m^2}{n^3}\right)^5$

26. $(-2x)^4$

27. $-(2x)^4$

28. $\left(-\frac{7}{x^3}\right)^2$

29. $\left(\frac{-3a^3}{5b^4}\right)^2$

30. Evaluate the expression $b^2 - 4ac$ for $a=4$, $b=-2$, and $c=-3$.

Comparing Addition and Multiplication

Add the like terms and simplify the products.

31. $2x+2x+2x$

32. $(2x)(2x)(2x)$

33. $3x^4+5x^4$

34. $(3x^4)(5x^4)$

35. If possible, add $2x^2+5x^3$

36. If possible, multiply $(2x^2)(5x^3)$

37. The formula for the area of a square is $A=x^2$ where x is the length of a side of the square.

(a) Complete the table of values for the area of a square with sides of length x cm.

Length of a side, x cm	**Area, cm^2**
1	
2	
3	
4	

(b) How does the area of a square with sides of length 2 cm compare to the area of a square with sides of length 4 cm?

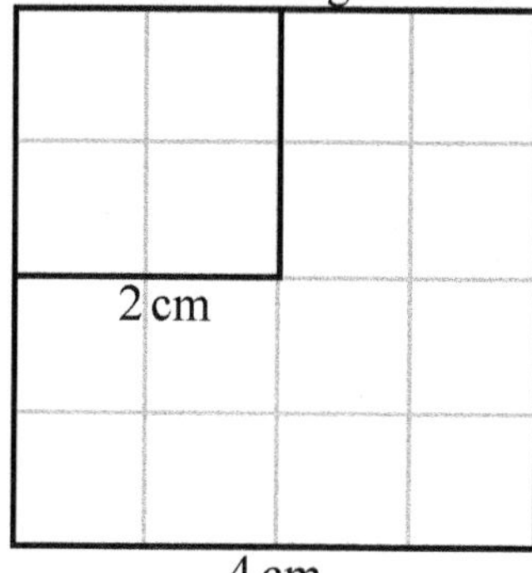

5.2 Lecture Guide: Quotient Rule and Zero Exponents

Objective 1: Use the quotient rule for exponents.

1. Complete the warm-up examples below. Assume $x \neq 0$.

Expanded Form:

$$\frac{x^5}{x^2} = \frac{x \cdot x \cdot x \cdot x \cdot x}{x \cdot x}$$

$$= \frac{x \cdot x \cdot x}{1}$$

$= \_\_\_\_\_\_$ (in exponential form)

Alternate form:

$$\frac{x^5}{x^2} = x^{5-2}$$

$= \_\_\_\_\_\_$

Quotient Rule for Exponents where $m > n$

Algebraically	**Verbally**	**Algebraic Example**
For any real number x and natural numbers m and n with $m > n$, $\frac{x^m}{x^n} = x^{m-n}$ for $x \neq 0$ $\frac{x^m}{x^n}$ is undefined for $x = 0$	To divide two expressions with the ________________ base, use the common base and ________________ the exponents.	$\frac{x^6}{x^2} = \_\_\_\_\_\_\_\_\_\_$

Simplify each expression.

2. $\frac{d^9}{d^3}$

3. $\frac{m^6 n^6}{m^4 n^3}$

4. $\frac{7^8}{7^6}$

5. $\frac{2x^{10}}{6x^4}$

6. $\frac{20x^{20}}{10x^{10}}$

7. $\frac{-12x^8 y^{14}}{2x^2 y^4}$

8. Complete the warm-up examples below. Assume $x \neq 0$.

Expanded Form:

$$\frac{x^2}{x^5} = \frac{x \cdot x}{x \cdot x \cdot x \cdot x \cdot x}$$

$$= \frac{1}{x \cdot x \cdot x}$$

$$= \_\_\_\_\_\_$$

Alternate form:

$$\frac{x^2}{x^5} = \frac{1}{x^{5-2}}$$

$$= \_\_\_\_\_\_$$

Quotient Rule for Exponents where $m < n$

The expression $\frac{x^m}{x^n}$ can also be simplified when $m < n$. The reasoning is similar to that when $m > n$. If $x \neq 0$ and m and n are natural numbers, then $\frac{x^m}{x^n} = \frac{1}{x^{n-m}}$ for $m < n$.

In the next section, we will examine another way to interpret $\frac{x^m}{x^n}$ when $m < n$.

Simplify each expression, assuming all bases are nonzero.

9. $\frac{m^2}{m^8}$

10. $\frac{x^4 y^2}{x^{12} y^6}$

11. $-\frac{6a^3b^8}{12a^9b^2}$

12. $\frac{50x^5y^4}{25x^{10}y^8}$

Objective 2: Simplify expressions with zero exponents.

13. Complete the warm-up examples below. Assume $x \neq 0$.

Expanded Form:

$$\frac{x^3}{x^3} = \frac{x \cdot x \cdot x}{x \cdot x \cdot x}$$

$$= x^{3-3}$$

$= \_\_\_\_\_\_$ (in exponential form)

Alternate form:

$\frac{x^3}{x^3} = \frac{\overset{1}{\cancel{x}} \cdot \overset{1}{\cancel{x}} \cdot \overset{1}{\cancel{x}}}{\underset{1}{\cancel{x}} \cdot \underset{1}{\cancel{x}} \cdot \underset{1}{\cancel{x}}} = 1$ (as a reduced fraction)

Zero Exponents:

Algebraically	**Verbally**	**Algebraic Example**
For any nonzero real number x, $x^0 = 1$. 0^0 is undefined.	Any ______________ real number raised to the 0 power equals 1.	$12^0 = \_\_\_\_\_\_\_\_\_\_\_\_$

Simplify each expression if possible.

14. **(a)** $0 \div 3$ **(b)** $3 \div 0$ **(c)** 0^3 **(d)** 3^0

Simplify each expression, assuming all bases are nonzero.

15. $(6a)^0$

16. $6a^0$

17. $(6a+b)^0$

18. $6a^0 + b^0$

Objective3: Combine the properties of exponents to simplify expressions.

Summary of the Properties of Exponents

For any nonzero real numbers x and y and whole number exponents m and n,

Product rule: $x^m \cdot x^n =$ ____________

Power rules: $(x^m)^n =$ _____ $(xy)^m =$ _____ $\left(\frac{x}{y}\right)^m =$ _____

Quotient rule: $\frac{x^m}{x^n} =$ ____________ for $m > n$

$\frac{x^m}{x^n} =$ ____________ for $m < n$

Zero exponent: $x^0 =$ ____________ for $x \neq 0$

0^0 is ____________

Simplify each expression. Assume all bases are nonzero.

19. $\left(\dfrac{2x^5}{x^3}\right)^4$

20. $\dfrac{28x^8y^{10}}{21x^{12}y^6}$

21. $\left[\left(3x^2\right)\left(5x^4\right)\right]^2$

22. $\left(\dfrac{24x^2y^8}{8x^6y^4}\right)^2$

23. $\left(\dfrac{36x^3}{12x^2}\right)\left(\dfrac{15x^7}{45x^4}\right)$

24. $\dfrac{\left(5x^3y^5\right)^2}{\left(3xy^3\right)^4}$

25. Use the formula $A = P(1+r)^t$ to find the value of a $15,000 investment compounded at 9% at the end of each year over a six-year period.

Year, t	Value, A
0	$15,000
1	
2	
3	
4	
5	
6	

Comparing Subtraction and Division

Subtract the like terms and simplify the quotients. Assume $x \neq 0$.

26. $12x^3 - 4x^3$

27. $\dfrac{12x^3}{4x^3}$

28. If possible, subtract $16x^3 - 4x$

29. If possible, divide $\dfrac{16x^3}{4x}$

5.3 Lecture Guide: Negative Exponents and Scientific Notation

Objective 1: Simplify expressions with negative exponents.

1. **(a)** In the previous section, we stated the quotient rule as $\frac{x^m}{x^n} = \frac{1}{x^{n-m}}$ for $x \neq 0$ and $n > m$.

Use this rule to simplify: $\frac{x^4}{x^7} =$ ____________

(b) Assume that the quotient rule, which states that $\frac{x^m}{x^n} = x^{m-n}$ for $x \neq 0$, is true for all integral values of *m* and *n*. Use this rule to simplify: $\frac{x^4}{x^7} =$ ____________

(c) Since we want the two expressions above to be equal we must have _________ = _________.

Generalizing, we get the following result for negative exponents:

Negative Exponents

Algebraically	**Verbally**	**Algebraic Example**
For any nonzero real number x and natural number n, $x^{-n} = \frac{1}{x^n}$.	A ______________ base with a negative exponent can be rewritten by using the ______________ of the base and the corresponding positive exponent.	$x^{-4} =$

Simplify each expression

2. 2^{-4}

3. -2^4

4. $3^{-1} + 5^{-1}$

5. $(3+5)^{-1}$

Note the effect of a negative exponent on a fraction.

6. Simplify $\left(\frac{2}{3}\right)^{-1} = \dfrac{1}{\frac{2}{3}} =$ ____________

Fraction to a Negative Power

Algebraically	**Verbally**	**Numerical Example**
For any nonzero real numbers x and y and natural number n, $\left(\frac{x}{y}\right)^{-n} = \left(\frac{y}{x}\right)^{n}$.	A nonzero fraction to a negative exponent can be rewritten by taking the _______________ of the fraction and using the corresponding positive exponent.	$\left(\frac{2}{7}\right)^{-2} =$

Simplify each expression.

7. $\left(\frac{2}{5}\right)^{-3}$

8. $\left(\frac{2}{3}\right)^{-2}$

9. $\dfrac{1}{3^{-2}}$

10. $\dfrac{2^{-2}}{3}$

11. $(5x)^{-2}$

12. $5x^{-2}$

13. $\dfrac{3x^{-2}}{y}$

14. $\left(\dfrac{-3x}{y}\right)^{-2}$

Objective 2: Use the properties of exponents to simplify expressions.

Summary of the Exponent Rules

For any nonzero real numbers x and y and whole number exponents m and n,

Product rule: $x^m \cdot x^n =$ ____________

Power rules: $(x^m)^n =$ _____ $(xy)^m =$ _____ $\left(\frac{x}{y}\right)^m =$ _____

Quotient rule: $\frac{x^m}{x^n} =$ ____________

Zero exponent: $x^0 =$ ____________ for $x \neq 0$

Negative exponent rule: $x^{-n} = \frac{1}{x^n}$

Simplify each expression to a form involving only positive exponents. Assume $x \neq 0$ and $y \neq 0$.

15. $\left(\frac{x^5}{x^{-3}}\right)^4$

16. $\left[\left(3x^{-2}\right)\left(2x^5\right)\right]^{-2}$

17. $\frac{x^2 y^{-3}}{x^{-1} y^4}$

18. $\frac{\left(5x^{-3}y^{-5}\right)^2}{\left(3xy^{-3}\right)^4}$

19. $\left(\dfrac{36x^{3}}{12x^{-2}}\right)\left(\dfrac{15x^{-7}}{45x^{4}}\right)$

20. $\left(\dfrac{14x^{-3}y^{2}}{35x^{2}y^{-4}}\right)^{-2}$

21. $\left(-2x^{-3}y^{4}\right)^{2}\left(4x^{3}y^{-6}\right)^{-1}$

22. $\dfrac{\left(10x^{-2}y^{4}\right)^{-2}\left(-2x^{6}y^{-1}\right)}{\left(25xy^{-4}\right)^{-1}}$

Objective 3: Use scientific notation.

Writing a Number in Standard Decimal Notation

Verbally	**Numerical Examples**
Multiply out the two factors by using the given power of ten.	
a. If the exponent on 10 is positive, move the decimal point to the ____________.	**a.** $3.456 \times 10^2 = 3.456 \times 100 = 345.6$ The decimal point is moved 2 places to the right.
b. If the exponent on 10 is ____________, do not move the decimal point.	**b.** $3.456 \times 10^0 = 3.456 \times 1 = 3.456$ The decimal point is not moved.
c. If the exponent on 10 is negative, move the decimal point to the ____________.	**c.** $3.456 \times 10^{-2} = 3.456 \times 0.01 = 0.03456$ The decimal point is moved 2 places to the left.

Write each number in standard decimal notation.

23. 5.71×10^4

24. 4.25×10^{-4}

25. 3.2×10^{-6}

26. 3.987×10^7

Writing a Number in Scientific Notation

Verbally	**Numerical Examples**
1. Move the decimal point immediately to the ____________ of the first nonzero digit of the number.	
2. Multiply by a power of 10 determined by counting the number of places the decimal point has been moved.	$3.456 = 3.456 \times 10^0$
a. The exponent on 10 is 0 or positive if the magnitude of the original number is 1 or ____________.	$345.6 = 3.456 \times 10^2$
b. The exponent on 10 is ____________ if the magnitude of the original number is less than 1.	$0.03456 = 3.456 \times 10^{-2}$

Write each number in scientific notation.

27. 80,000

28. 72,300

29. 0.008

30. 0.0000985

31. Write the result on the calculator screen in scientific notation and in standard decimal notation. See Technology Perspective 5.3.1.

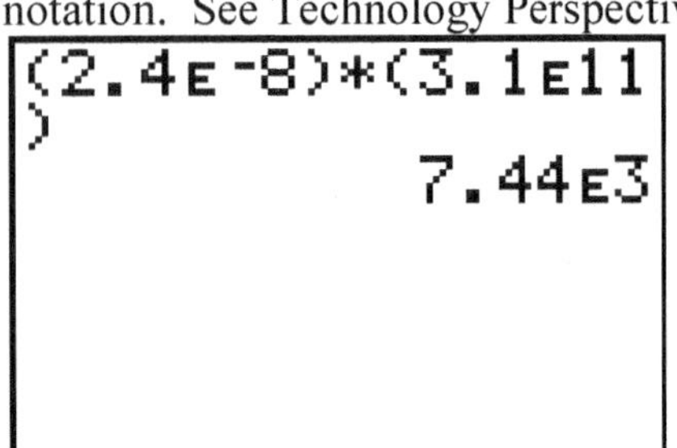

Scientific notation:

Standard decimal notation:

32. Each song on a personal music player requires about 4×10^6 bytes of memory. If the music player has 80 GB (8.0×10^{10} bytes) of memory available, approximate the number of songs it will hold.

33. Use scientific notation to estimate (4,990,000)(0.000147).

Pencil and Paper Estimate:

Calculator Approximation:

Evaluate each expression for $x = 2$ and $y = -3$.

34. $-x^{-2}+y^{-2}$

35. $-(x+y)^2$

36. $(x+y)^{-2}$

5.4 Lecture Guide: Adding and Subtracting Polynomials

Objective 1: Use the terminology associated with polynomials.

Monomials and Polynomials

	Verbally	**Algebraic Examples**
Monomials	A monomial is a real number, a variable, or a product of real numbers and variables with ________________ number exponents.	-5, π, x, A, $5x$, $7xy$ and πr^2 are monomials.
Polynomials	A polynomial is a monomial or a sum of a finite number of ________________.	-5 and $7xy$ are polynomials. $7x^2+9xy-3y^2$ is a polynomial.
Binomial	A binomial is a polynomial containing ________________ terms.	$3x-4$
Trinomial	A trinomial is a polynomial containing ________________ terms.	$2x^2+5x-7$

Degree of a Monomial

The degree of a monomial is the sum of the ________________ for all the variables in this term. A nonzero constant is understood to have degree ________________ $(4 = 4x^0$ with exponent $0)$, but no degree is assigned to the monomial 0.

1. Determine whether each expression is a monomial. If the expression is a monomial, give its coefficient and the degree of each monomial.

(a) $-3x^4$

Monomial? **Yes / No**

Coefficient: ______

Degree: ______

(b) $3x^{-4}$

Monomial? **Yes / No**

Coefficient: ______

Degree: ______

(c) π

Monomial? **Yes / No**

Coefficient: ______

Degree: ______

(d) xyz

Monomial? **Yes / No**

Coefficient: ______

Degree: ______

2. Determine whether each expression is a polynomial. Classify each polynomial according to the number of terms it contains.

(a) $4x+7$

Polynomial? **Yes / No**

Classification: ________________

(b) $\frac{9x}{2}$

Polynomial? **Yes / No**

Classification: ________________

(c) $\frac{2}{9x}$

Polynomial? **Yes / No**

Classification: ________________

(d) $12x^3+x+4$

Polynomial? **Yes / No**

Classification: ________________

Standard Form of a Polynomial

A polynomial is in standard form if (1) the variables in each term are written in ________________order and (2) the terms are arranged in ________________ powers of the first variable.

Degree of a Polynomial

The degree of a polynomial is the same as the degree of the term with the ________________ degree. To find this highest degree, examine each term individually---do *not* sum the degrees of the terms.

3. Write each polynomial in standard form, and give the degree of each polynomial.

(a) $-9+3x^2+8x$

Standard form:

Degree:

(b) $-9y^2zx^5$

Standard form:

Degree:

(c) $-4v+9v^5+3v^2-v^3+1$

Standard form:

Degree:

Objective 2: Add and subtract polynomials.
When adding or subtracting polynomials, remember to use the ____________________ property to remove any grouping symbols and then add like terms.

Remember that **like terms** have exactly the same ________________ factors.

Identify each pair of terms as like or unlike.

4. $3x^2y$ and $3xy^2$

5. $-2x^3y$ and $11x^3y$

Determine each sum or difference and write the result in standard form.

6. $(8x-5)+(2x+9)$

7. $(4x^2+2x)-(2x^2-6x)$

Determine each sum or difference and write the result in standard form.

8. $(3x^2+3x+5)+(2x^2-3x-10)$

9. $(5x^2+3x+2)-(x^2-2x+7)$

10. $3(x^2+3x+5)+2(x^2-3x-10)$

11. $2(3x^2+4x-5)-3(4x^2-5x+1)$

12. $(x^2-3x+7)+3(x^2-4x+2)-(2x^2-1)$

13. $(x^2+2xy-6y^2+1)-(5x^2-4xy+7y^2+2)$

14. The profit in dollars made by selling x units is given by the polynomial $P(x) = -x^2 + 25x - 46$. Evaluate and interpret each expression.

(a) Evaluate $P(0)$

Interpret $P(0)$

(b) Evaluate $P(2)$

Interpret $P(2)$

(c) Evaluate $P(13)$

Interpret $P(13)$

15. Write a polynomial for the perimeter of the polygon.

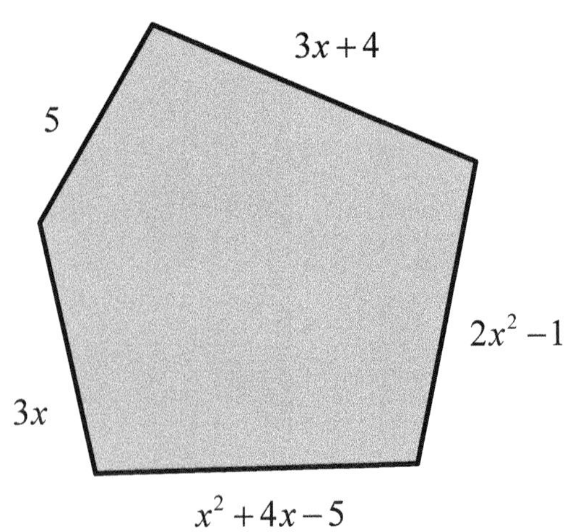

5.5 Lecture Guide: Multiplying Polynomials

Objective 1: Multiply polynomials.

Multiplying a Monomial Times a Polynomial

Verbally	**Algebraic Example**
To multiply a monomial times a polynomial, use the distributive property to multiply the monomial times each _______________ of the polynomial.	$5x(2x^2-3x+8)=5x(2x^2)-5x(3x)+5x(8)$ $=10x^3-15x^2+40x$

Multiply the polynomial factors.

1. $(-5ab^2c^3)(-4a^2bc^2)$

2. $4x^2(5x+3)$

3. $-2x^2(2x^2-4xy-5y^2)$

4. $(8x^2-3xy-7y^2)(3y)$

Multiplying Two Polynomials

To multiply one polynomial times another, multiply each _______________ of the first polynomial times each _______________ of the second polynomial and then combine the like terms.

Multiply the polynomial factors.

5. $(x-3)(x+7)$

6. $(x-5)(x^2+2x+7)$

7. $(2x+5y)(2x-5y)$

8. $(x-5)(x^2+3x+3)$

9. $(x^2+2x-1)(x^2-3x+3)$

10. $(x-3)(x+2)(x-5)$

Objective 2: Multiply binomials by inspection.
A helpful method for multiplying two binomials is sometimes known as the FOIL method. **FOIL** is an acronym for **F**irst, **O**uter, **I**nner, and **L**ast. The logic of the FOIL method is illustrated below for the product $(2x+5)(x-6)$.

$$(2x+5)(x-6) = 2x(x-6)+5(x-6)$$
$$= 2x^2 - 12x + 5x - 30$$

Using the distributive property results in multiplying each term of $2x+5$ times each term of $x-6$.

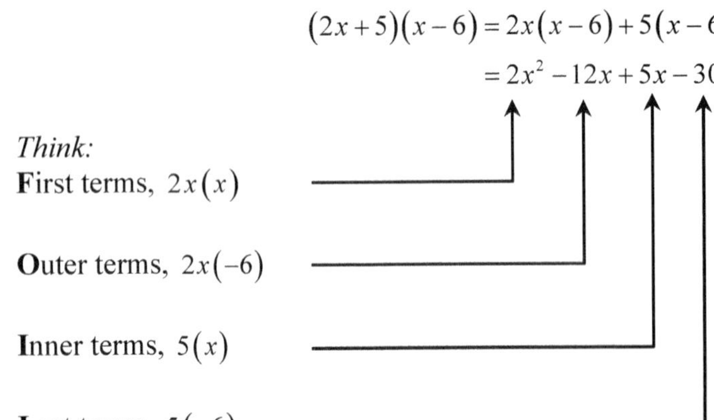

This product would be simplified to $2x^2 - 7x - 30$

Practice using the FOIL method to multiply each pair of binomials as quickly as possible.

11. $(x+6)(x+8)$

12. $(x+3)(x-5)$

13. $(x-4)(x-7)$

14. $(x-9)(x+2)$

15. $(2x-5)(3x+1)$

16. $(4x+3)(5x-2)$

17. $(8x+7)(2x+3)$

18. $(4x-5)^2$

19. Write a polynomial for the area of the figure.

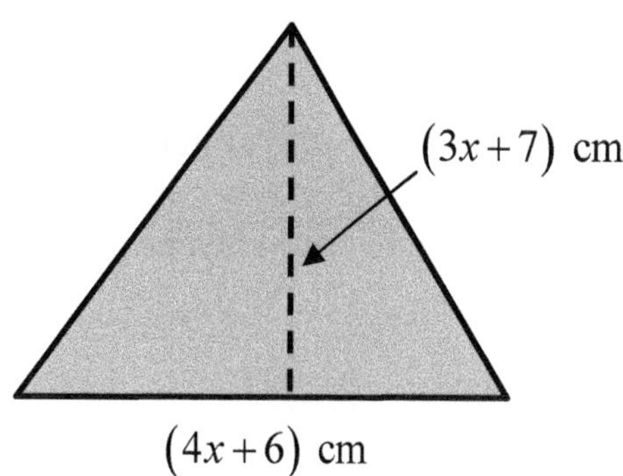

Use the distributive property to expand each expression in the first column and to factor each expression in the second column.

Expand	**Factor**
20. $2x(x^2-3x)$	**21.** $8x^2+12x$
22. $3xy(4x-5)$	**23.** $15x^2y+25xy$
24. $x(x+7)-2(x+7)$	**25.** $x(x+7)-2(x+7)$

26. A concrete sidewalk of width 5 ft is poured around a circular pool of radius x ft. From the formula $A=\pi r^2$ the area covered by the concrete will be $A=\pi(x+5)^2-\pi x^2$. Expand and simplify this polynomial.

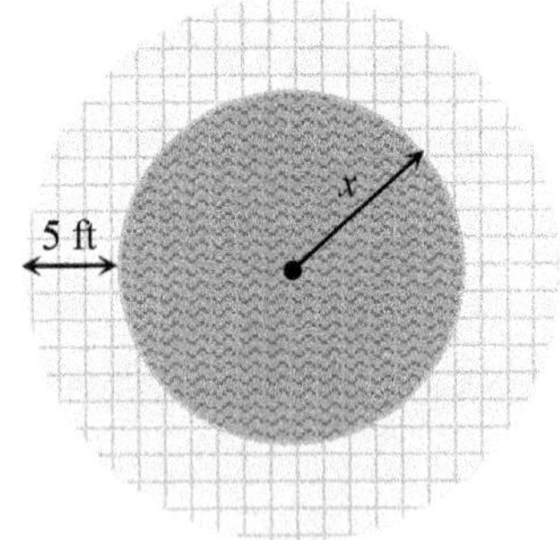

27. Revenue = Number of items sold • Price per item The price per unit is x and the number of units that will sell at that price is $5000-8.75x$.

(a) Write a revenue polynomial $R(x)$ for this product.

(b) Evaluate and interpret $R(52)$.

5.6 Lecture Guide: Special Products of Binomials

Objective 1: Determine the product of a sum and a difference of binomials by inspection.

Multiply each pair of factors. See if you can observe a pattern as you form the products.

Factors	**Products**
Example: $(x-1)(x+1)=x^2+x-x-1$	$=x^2-1$
1. $(3x+2)(3x-2)$	
2. $(4x+3)(4x-3)$	
3. $(7x-4)(7x+4)$	
4. $(3x+5)(3x-5)$	

5. Can you describe what you observed?

Sum Times a Difference

Algebraically	**Verbally**	**Algebraic Example**
$(A+B)(A-B)=A^2-B^2$	This product of a sum times a difference is the ______________ of their squares.	$(x+9)(x-9)=$

Use the pattern for a sum times a difference to multiply each pair of factors.

Factors	**Products**
Example: $(x+6)(x-6)=x^2-6^2$	$=x^2-36$
6. $(x-7)(x+7)$	
7. $(5x+8)(5x-8)$	
8. $(2x-9)(2x+9)$	
9. $(3x-10)(3x+10)$	
10. $(w+12)(w-12)$	
11. $(8v-1)(8v+1)$	
12. $(2a+3b)(2a-3b)$	
13. $(3x-5)(3x+5)$	
14. $(5x+7y)(5x-7y)$	

Objective 2: Determine the square of a binomial by inspection.

Multiply each pair of factors. See if you can observe a pattern as you form the products.

Factors	**Products**
Example: $(x+1)^2 = (x+1)(x+1) = x^2 + x + x + 1$	$= x^2 + 2x + 1$
15. $(3x+2)^2 = (3x+2)(3x+2)$	
16. $(5x-3)^2 = (5x-3)(5x-3)$	
17. $(3x-4)^2 = (3x-4)(3x-4)$	
18. $(6x+5)^2 = (6x+5)(6x+5)$	

Square of a Sum and Square of a Difference

Algebraically	**Verbally**	**Algebraic Example**
Square of a Sum $(A+B)^2 = A^2 + 2AB + B^2$ **Square of a Difference** $(A-B)^2 = A^2 - 2AB + B^2$	A trinomial that is the square of a binomial has: **1.** a first term that is a ____________ of the first term of the binomial. **2.** a middle term that is ____________ the product of the two terms of the binomial. **3.** a last term that is a ____________ of the last term of the binomial.	$(x+1)^2 = x^2 + 2x + 1$ $(x-1)^2 = x^2 - 2x + 1$

Use the patterns for a square of a sum and the square of a difference to multiply each pair of factors.

Factors	**Products**
Example: $(x+6)^2 = x^2 + 2(6x) + 6^2$	$= x^2 + 12x + 36$
19. $(x-7)^2$	
20. $(5x+8)^2$	
21. $(2x-9)^2$	
22. $(3x-10)^2$	
23. $(7m+n)^2$	
24. $(3a-4b)^2$	
25. $(5x+6y)^2$	

One reason for practicing the multiplication of special products is to recognize these products when factoring.

Select the correct factored form for each polynomial to complete each statement.

26. $x^2 - 10x + 25 =$ ________________

27. $x^2 - 25 =$ ________________

28. $4x^2 + 20x + 25 =$ ________________

29. $4x^2 - 25 =$ ________________

A. $(x+5)^2$

B. $(x-5)^2$

C. $(2x+5)^2$

D. $(2x-5)^2$

E. $(x+5)(x-5)$

F. $(2x+5)(2x-5)$

Objective 3: Use the order of operations to simplify polynomials.

Order of Operations

Step 1: Start with the expression within the innermost pair of grouping symbols.

Step 2: Perform all exponentiations.

Step 3: Perform all multiplications and divisions as they appear from left to right.

Step 4: Perform all additions and subtractions as they appear from left to right.

Use the correct order of operations to simplify each polynomial.

30. **(a)** $3x - 7(3x+7)$ **(b)** $(3x-7)(3x+7)$

31. **(a)** $(3x+7)^2$ **(b)** $(3x)^2 + (7)^2$

32. $(2x+9)(2x-9)-(2x-9)^2$

33. $2(3x-4y)^2-5(3x+4y)^2$

34. Subtract the square of $x-4$ from the square of $x+2$.

35. Expand: $(3x+4)(3x-4)$ **Factor:** $9x^2-16$

36. Expand: $(3x+4y)^2$ **Factor:** $9x^2+24xy+16y^2$

5.7 Lecture Guide: Dividing Polynomials

Objective 1: Divide a polynomial by a monomial.

Find each quotient and assume the variables are restricted to values that avoid division by zero.

1. $\dfrac{18x^5y^4}{3x^2y^2}$

2. $\dfrac{16x^5y^3}{40x^2y^6}$

Dividing a Polynomial by a Monomial

Verbally	Algebraic Example
To divide a polynomial by a monomial, divide each ____________ of the polynomial by the monomial.	$\dfrac{6x^2+4x}{2x}=\dfrac{6x^2}{2x}+\dfrac{4x}{2x}$ $=3x+2$

Find each quotient and assume the variables are restricted to values that avoid division by zero.

3. $\dfrac{21x^4-6x^3y-9x^2y^2}{3x^2}$

4. $\dfrac{-28x^2y+8xy^2-12y^3}{4y}$

5. $\dfrac{12x^4y^2+9x^3y^3-3x^3y^2}{3x^3y^2}$

6. $\dfrac{-28x^3y+8x^2y^2-12xy^3}{4x^2y}$

Objective 2: Use long division of polynomials.

Long Division of Polynomials

Step 1.	Write the polynomials in long-division format, expressing each in _______________ form.
Step 2.	Divide the first term of the divisor into the first term of the dividend. The result is the first term of the _______________.
Step 3.	Multiply the first term of the quotient times every term in the divisor, and write this product under the dividend, aligning like terms.
Step 4.	_______________ this product from the dividend, and bring down the next term.
Step 5.	Use the result of Step 4 as a new dividend, and repeat Steps 2-4 until either the remainder is zero or the degree of the remainder is less than the degree of the divisor.

7. Find the quotient $4x-5\overline{)8x^2+2x-15}$.

8. Check the answer in problem 7 by multiplying the divisor by the quotient.

Find each quotient.

9. $5x+4\overline{)10x^2-17x-20}$

10. $6x+5\overline{)18x^2-9x-20}$

Find each quotient.

11. $\dfrac{x^3-12x-16}{x+2}$

Hint: Watch out for the missing term.

12. $\dfrac{3x^2+4x-10}{x-2}$

Hint: This will have a non-zero remainder.

13. The area of a triangle is $\left(4x^2-4x-35\right)\text{mm}^2$. The base of the triangle is $(2x-7)\,\text{mm}$. Find the altitude of the triangle. (Hint: $A=\frac{1}{2}bh$.)

14. Using x to represent the number of units produced, a factory determined that the cost in dollars to produce these units was $C(x)=150x+540$.

(a) Write an expression for $A(x)$, the average cost of producing x units.

(b) Evaluate and interpret $C(10)$.

(c) Evaluate and interpret $A(10)$.

15. (a) One factor of $6x^3+11x^2-17x-30$ is $3x-5$. Determine the other factor.

(b) Another factor of $6x^3+11x^2-17x-30$ is $2x+3$.
Complete this equation: $6x^3+11x^2-17x-30=(2x+3)(3x-5)(\_\_\_\_\_\_\_\_\_)$
Hint: Divide $2x+3$ into the quotient from part **(a)**.

16. When a polynomial is divided by $3x-4$, the quotient is $x+3+\frac{10}{3x-4}$. Determine this polynomial.

Lecture Guides for

Chapter 6

Factoring Polynomials

6.1 Lecture Guide: An Introduction to Factoring Polynomials

Objective 1: Factor the GCF out of a polynomial.

Greatest Common Factor of a Polynomial

The GCF of a polynomial is the common factor that contains
1. the largest possible numerical coefficient and
2. the largest possible exponent on each variable factor.

Determine the GCF of each pair of monomials.

1. $24x^3y^2$ and $36xy^2$

2. $15x^2y$ and $25xy$

Factor out the GCF of each polynomial.

3. $24x^3y^2 - 36xy^2$

4. $15x^2y + 25xy$

5. $15x^2y + 45x^2y^3 - 30xy^3$

6. $a(3a+2b)+5b(3a+2b)$

7. **(a)** Factor out the GCF of the polynomial $x(x-1)+4(x-1)$.

(b) Use your calculator to complete the table below by letting Y_1 equal the original polynomial and Y_2 equal the factored form.

x	$Y_1 = x(x-1)+4(x-1)$	$Y_2 =$
−3		
−2		
−1		
0		
1		
2		
3		

(c) Graph Y_1 and Y_2 in the standard viewing window. How do the graphs compare?

(d) What would you conclude about the original polynomial $x(x-1)+4(x-1)$ and the factored form?

Objective 2: Factor by Grouping.

Factoring a Four-Term Polynomial by Grouping Pairs of Terms

Step 1. Be sure you have factored out the GCF if it is not 1.	Example: $ax + 2ay + 2bx + 4by$
Step 2. Use grouping symbols to pair the terms so that each pair has a common factor other than 1.	
Step 3. Factor the GCF out of each pair of terms.	
Step 4. If there is a common binomial factor of these two groups, factor out this GCF. If there is no common binomial factor, try to use Step 2 again with a different pairing of terms. If all possible pairs fail to produce a common binomial factor, the polynomial will not factor by this method.	

Factor each polynomial by the grouping method.

8. $ax + 3a + 8bx + 24b$

9. $2x^2 + 16x + 3x + 24$

10. $3x^2 - 9x + 2x - 6$

11. $6x^2 - 8x - 15x + 20$

Objective 3: Use the zeros of a polynomial $P(x)$ and the x-intercepts of the graph of $y = P(x)$ to factor the polynomial.

The relationship among the factors of a polynomial, the zeros of a polynomial function, and the x-intercepts of a graph of a polynomial function is an important one.

If $x = c$ is an input value for which the output $P(c)$ equals 0, then c is called a **zero of the function**.

Equivalent Statements about Linear Factors of a Polynomial

For a real constant c and a real polynomial $P(x)$, the following statements are equivalent.

Graphically	**Numerically**	**Algebraically**
$(c, 0)$ is an x-intercept of the graph of $y = P(x)$	$P(c) = 0$, that is, c is a zero of $P(x)$	$x - c$ is a factor of $P(x)$

12. Consider the polynomial $P(x) = (x-1)(x+5)$. Use the factored form of $P(x)$ to evaluate each expression.

(a) $P(-5)$

(b) $P(-3)$

(c) $P(-1)$

(d) $P(1)$

13. Use the table and graph provided for $P(x) = (x-1)(x+5)$ to answer each question.

x	$P(x)$
-5	0
-4	-5
-3	-8
-2	-9
-1	-8
0	-5
1	0

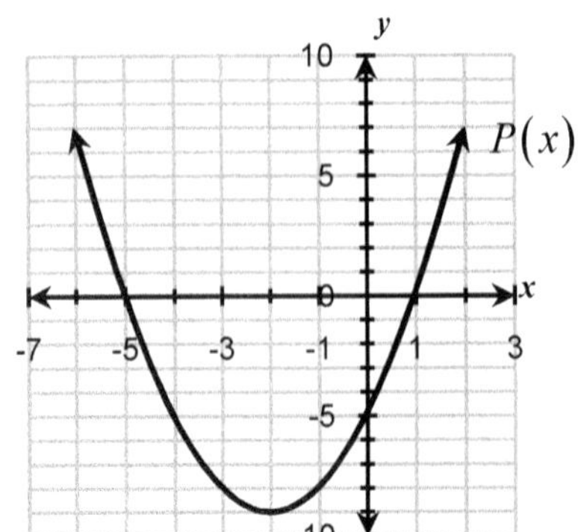

(a) List the factors of $P(x)$.

(b) List the zeros of $P(x)$.

(c) List the x-intercepts of the graph.

14. Consider the polynomial $P(x)=x^2+2x-3$.

x	$P(x)$
−4	5
−3	0
−2	−3
−1	−4
0	−3
1	0
2	5

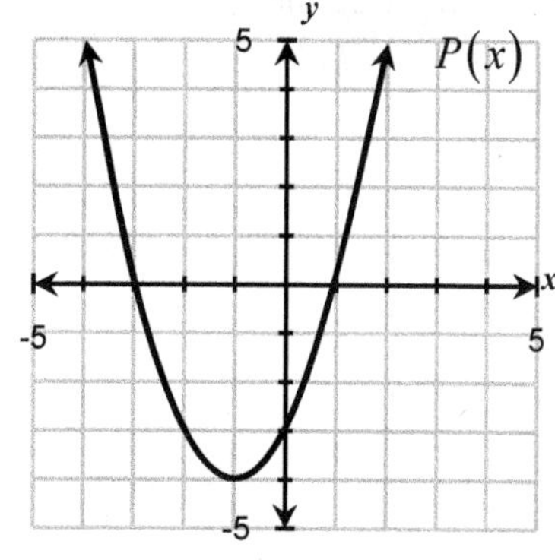

(a) List the zeros of $P(x)$.

(b) List the x-intercepts of the graph.

(c) Use parts **(a)** and **(b)** to determine the factored form of the polynomial $P(x)=x^2+2x-3$.

Use the factored form of each polynomial to list the zeros and the x-intercepts of the graph of $y=P(x)$.

15. $P(x)=(x-3)(x-4)$

Zeros:

***x*-intercepts:**

16. $P(x)=(x+5)(x+4)(x-2)$

Zeros:

***x*-intercepts:**

Complete the following table for each polynomial.

Polynomial $P(x)$	**Factored Form**	**Zeros of** $P(x)$	***x*-intercepts of the graph of** $y=P(x)$
17. $x^2+3x-54$	$(x-6)(x+9)$		
18. $x^2-18x+80$		8 and 10	
19. x^2-5x-6			$(-1,0),(6,0)$
20. x^3-2x^2-63x		−7, 0, and 9	

21. Use the given graph for $P(x)=x^2+2x-8$ to factor this polynomial.

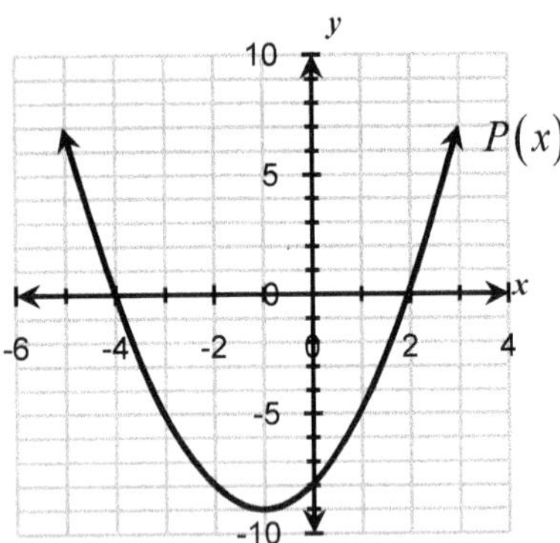

22. Use the given table for $P(x)=x^3-8x^2+12x$ to factor this polynomial.

x	$P(x)$
0	0
1	5
2	0
3	−9
4	−16
5	−15
6	0

23. Use a graphing calculator to create a graph for $P(x)=x^3-3x^2-10x+24$. Then use the zeros to factor the polynomial.

Sketch:

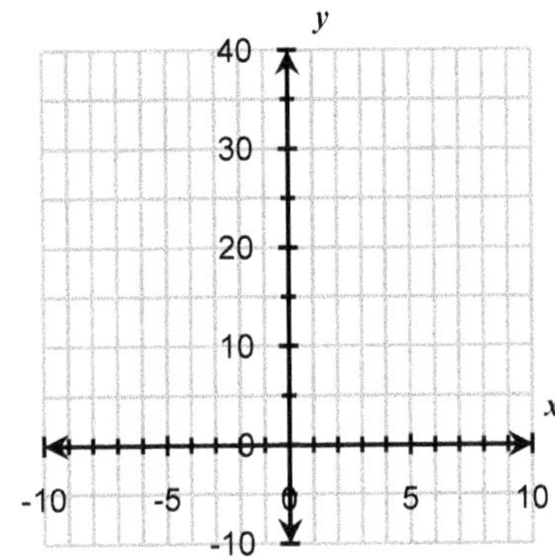

Zeros:

Factored form:

24. Use a graphing calculator or a spreadsheet to complete the table for $P(x)=x^4-10x^2+9$. Then use the zeros to factor the polynomial.

x	$P(x)$
−3	
−2	
−1	
0	
1	
2	
3	

Zeros:

Factored form:

Factoring a Polynomial with Two Variables Use the factored form of a polynomial in one variable to factor each polynomial in two variables.

25. Given: $x^2-14x+24=(x-2)(x-12)$

Factor: $x^2-14xy+24y^2$

26. Given: $x^2-5x-36=(x+4)(x-9)$

Factor: $x^2-5xy-36y^2$

6.2 Lecture Guide: Factoring Trinomials of the Form $x^2 + bx + c$.

Objective 1: Factor trinomials of the form $x^2 + bx + c$ by the trial-and-error method or by inspection.

A trinomial in the form $ax^2 + bx + c$ is called a quadratic trinomial. Your goal for this section should be to be able to factor quadratic trinomials as quickly and efficiently as possible. We will restrict our focus to the case where the leading coefficient $a = 1$ for this section.

To begin understanding the procedure we use to factor these trinomials, examine the following three products.

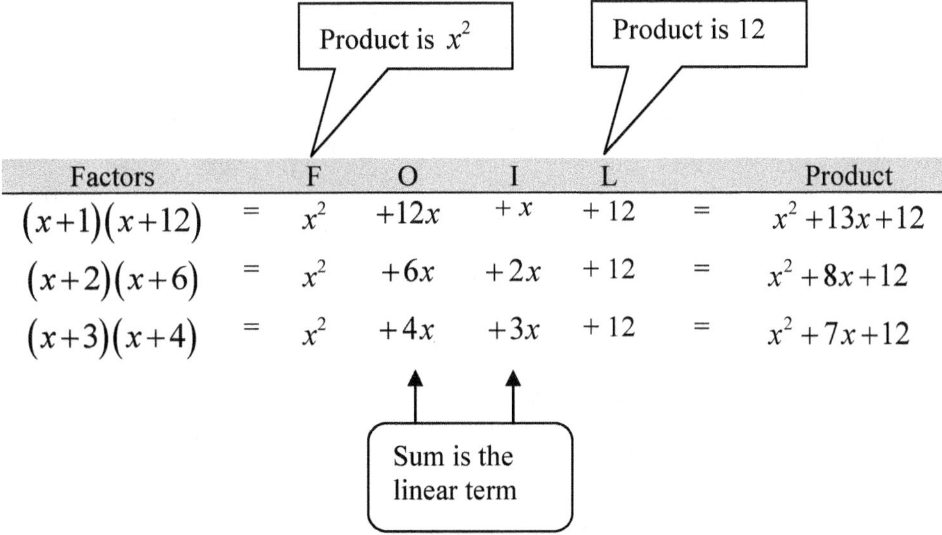

Factors		F	O	I	L		Product
$(x+1)(x+12)$	=	x^2	$+12x$	$+x$	$+12$	=	$x^2+13x+12$
$(x+2)(x+6)$	=	x^2	$+6x$	$+2x$	$+12$	=	$x^2+8x+12$
$(x+3)(x+4)$	=	x^2	$+4x$	$+3x$	$+12$	=	$x^2+7x+12$

Thus, a factorization of $x^2 + bx + c$ must have a pair of factors of c whose sum is b.

Factorable Trinomials

Algebraically	**Verbally**	**Example**
$x^2+bx+c=(x+c_1)(x+c_2)$ where $c_1c_2=c$ and $c_1+c_2=b$.	A trinomial x^2+bx+c is factorable into a pair of binomial factors with integer coefficients if and only if there are two integers whose ____________ is c and whose ____________ is b. Otherwise, the trinomial is prime over the integers.	$x^2+14x+24=(x+2)(x+12)$ where $2\cdot 12=$ ______ and $2+12=$ ______. $x^2+50x+24$ is prime because there are no factors of ______ whose sum is ______.

Factoring by Trial-and-Error or by Inspection

To factor $x^2 + bx + c$ by trial and error list the possible factors of c and examine them until you find the right combination that produces the correct linear term. To factor $x^2 + bx + c$ by inspection, try to mentally determine two factors of c whose sum is b. Initially, it is a good idea to list the factors. This will help you to notice patterns and develop a sense for factoring trinomials. With practice you will be able to factor many of the trinomials in this section by inspection.

In problems 1-8, the values below the binomial factors represent the possible factors of the constant term c in the quadratic trinomial. Use this information along with the given sign pattern to determine the factored form.

1. $x^2 + 21x + 20 = (x + \quad)(x + \quad)$

1	20
2	10
4	5

2. $x^2 + 9x + 20 = (x + \quad)(x + \quad)$

1	20
2	10
4	5

3. $x^2 - 11x + 30 = (x - \quad)(x - \quad)$

1	30
2	15
3	10
5	6

4. $x^2 - 13x + 30 = (x - \quad)(x - \quad)$

1	30
2	15
3	10
5	6

5. $x^2 + 4x - 12 = (x - \quad)(x + \quad)$

1	12
2	6
3	4

6. $x^2 + x - 12 = (x - \quad)(x + \quad)$

1	12
2	6
3	4

7. $x^2 - 16x - 36 = (x + \quad)(x - \quad)$

1	36
2	18
3	12
4	9
6	6

8. $x^2 - 5x - 36 = (x - \quad)(x + \quad)$

1	36
2	18
3	12
4	9
6	6

Note: What is wrong with problem 8? How can you fix it?

Sign Pattern for the Factors of $x^2 + bx + c$

Verbally	Algebraically	Examples
If the constant term is ____________, the factors of this term must have the same sign. These factors will share the same sign as the linear term.	$x^2+bx+c=(x+?)(x+?)$ $x^2-bx+c=(x-?)(x-?)$	$x^2+5x+6=(x+2)(x+3)$ $x^2-5x+6=(x-2)(x-3)$
If the constant term is ____________, the factors of this term must be opposite in sign. The sign of the constant factor with the larger absolute value will be the same as that of the linear term.	$x^2+bx-c=(x+?)(x-?)$ $x^2-bx-c=(x+?)(x-?)$	$x^2+5x-6=(x+6)(x-1)$ $x^2-5x-6=(x+1)(x-6)$

Match each trinomial with the appropriate sign pattern for the factors of this trinomial.

9. x^2+5x+6
10. x^2-5x+6
11. x^2-5x-6
12. x^2+5x-6

A. $(x+\_\_\_)(x+\_\_\_)$
B. $(x+\_\_\_)(x-\_\_\_)$
C. $(x-\_\_\_)(x-\_\_\_)$

Factor each trinomial by trial-and-error or by inspection.

13. $x^2+8x+12$

14. $x^2+8x+15$

15. $m^2-10m+21$

16. $t^2-10t+24$

Factor each trinomial by trial-and-error or by inspection.

17. $y^2 - 8y + 7$

18. $a^2 - 14a - 15$

19. $x^2 - x - 12$

20. $x^2 - 6x - 16$

21. $z^2 - 3z - 28$

22. $x^2 + 12x - 28$

23. $c^2 - 12c + 20$

24. $x^2 - 3x - 40$

Objective 2: Identify a prime trinomial of the form $x^2 + bx + c$.

A Prime Polynomial

A polynomial is **prime over the integers** if its only factorization must involve either ____ or _____ as one of the factors.

Use the given tables to assist you in identifying which trinomials are prime and factoring the trinomials that can be factored over the integers. Note that if we list all possibilities in a table and none of them work, then the polynomial must be prime.

Factors of 36		Sum of Factors
1	36	37
2	18	20
3	12	15
4	9	13
6	6	12

25. $x^2 + 15x + 36$

26. $x^2 + 14x + 36$

Factors of 24		Sum of Factors
−1	−24	−25
−2	−12	−14
−3	−8	−11
−4	−6	−10

27. $x^2 - 14x + 24$

28. $x^2 - 13x + 24$

Factors of −56		Sum of Factors
−1	56	55
−2	28	26
−4	14	10
−7	8	1

29. $x^2 + 20x - 56$

30. $x^2 + 10x - 56$

Objective 3: Factor trinomials of the form $x^2+bxy+cy^2$ by the trial-and-error method or by inspection.

Factor each trinomial by trial-and-error or by inspection.

31. $x^2+22xy-48y^2$

32. $x^2-5xy-36y^2$

33. $x^2+11xy+24y^2$

34. $x^2-8xy+12y^2$

35. $x^2-14xy+48y^2$

36. $x^2+11xy+10y^2$

Each of the following trinomials has a greatest common factor. First factor out the GCF and then complete the factorization.

37. $2x^2-2x-12$

38. $5x^2+45x+100$

39. $10x^2-20x-30$

40. $-x^2+4x+12$

41. $-x^2-9x-18$

42. $-3x^2+15x+42$

Each of the following trinomials has a greatest common factor. First factor out the GCF and then complete the factorization.

43. $-2x^2 + 12x - 10$

44. $3x^3y + 9x^2y^2 - 30xy^3$

45. $2x^3y^3 + 8x^2y^4 - 90xy^5$

46. $(a+2b)x^2 - 5(a+2b)x - 36(a+2b)$

Complete the following table for each polynomial.

Polynomial $P(x)$	**Factored Form**	**Zeros of** $P(x)$	***x*-intercepts of the graph of** $y = P(x)$
47. $x^2 + 3x - 18$			
48. $x^2 - 10x + 21$			

6.3 Lecture Guide: Factoring Trinomials of the Form $ax^2 + bx + c$.

Objective 1: Factor trinomials of the form $ax^2 + bx + c$ by the trial-and-error method.

Fill in the missing information to complete the factorization of each trinomial.

1. $2x^2 + 7x + 6 = (2x + \quad)(x + \quad)$

2. $2x^2 + 13x + 6 = (2x + \quad)(x + \quad)$

3. $3x^2 - 11x + 6 = (3x - \quad)(x - \quad)$

4. $3x^2 - 19x + 6 = (3x - \quad)(x - \quad)$

5. $3x^2 - x - 4 = (3x - \quad)(x + \quad)$

6. $3x^2 + 4x - 4 = (3x - \quad)(x + \quad)$

7. $5x^2 - 17x - 12 = (5x + \quad)(x - \quad)$

8. $5x^2 - 4x - 12 = (5x + \quad)(x - \quad)$

Fill in the missing + or – symbols to complete the factorization of each trinomial.

9. $2x^2 + 7x + 5 = (2x\_\_\_\_5)(x\_\_\_\_1)$

10. $3x^2 - 8x + 4 = (3x\_\_\_\_2)(x\_\_\_\_2)$

11. $4x^2 + 8x - 5 = (2x\_\_\_\_1)(2x\_\_\_\_5)$

12. $5x^2 - 7x - 6 = (5x\_\_\_\_3)(x\_\_\_\_2)$

It is very useful to be able to recognize patterns and be able to use these patterns to quickly factor some polynomials by inspection. Factor each trinomial by inspection.

13. $2x^2 + 7x + 3$

14. $2x^2 + 11x + 5$

15. $2x^2 + x - 3$

16. $2x^2 - 9x - 5$

17. $2x^2 - 5x - 3$

18. $2x^2 + 3x - 5$

Objective 2: Factor trinomials of the form $ax^2 + bx + c$ ***by the AC method.***
Objective 3: Identify a prime trinomial of the form $ax^2 + bx + c$.

Factoring a polynomial can be considered a reversal of the process of multiplying the factors of the polynomial. In Section 6.2, we focused on factoring trinomials where the leading coefficient was 1. Factoring trinomials where the leading coefficient is not 1 can be more complicated. We will start by multiplying several pairs of factors that form a trinomial with a leading coefficient of 6.

19. Multiply the factors in this table by writing out both middle terms and then simplify the result. The first row has been completed.

Factors	Products (F O I L)	
$(x+1)(6x+15)=$	$6x^2 + 15x + 6x + 15=$	$6x^2+21x+15$
$(x+3)(6x+5)=$	$6x^2 \qquad +15=$	
$(x+5)(6x+3)=$	$6x^2 \qquad +15=$	
$(x+15)(6x+1)=$	$6x^2 \qquad +15=$	
$(2x+1)(3x+15)=$	$6x^2 \qquad +15=$	
$(2x+3)(3x+5)=$	$6x^2 \qquad +15=$	
$(2x+5)(3x+3)=$	$6x^2 \qquad +15=$	
$(2x+15)(3x+1)=$	$6x^2 \qquad +15=$	

20. Answer each question about the table above.

(a) What is the product of the coefficients of each pair of middle terms?

(b) What do you notice about the first and last term of each product?

(c) What is the product of the coefficients of the first and last terms?

(d) What is the correct factorization of $6x^2+19x+15$?

(e) The procedure for factoring trinomials of the form ax^2+bx+c by the AC method involves finding two factors of *ac* whose sum is *b*. When expanded, the correct factorization of $6x^2+19x+15$ has two middle terms whose coefficients have a product of ____________ and a sum of ____________.

Factoring $ax^2 + bx + c$ by the AC Method

Procedure

Step 1. Factor out the GCF. If a is negative, factor out -1.

Step 2. Find a pair of factors of ac whose sum is b. If there is not a pair of factors whose sum is b, the trinomial is prime over the integers.

- If the constant c is positive, the factors of ac must have the ______________ ______________. These factors will share the same sign as the linear coefficient b.
- If the constant c is negative, the factors of ac must be ______________ in sign. The sign of b will determine the sign of the factor with the larger absolute value.

Step 3. Rewrite the linear term of $ax^2 + bx + c$ so that b is the sum of the factors from Step 2.

Step 4. Factor the polynomial from Step 3 by grouping the terms and factoring the GCF out of each pair of terms.

Example

Factor $6x^2 + 7x - 20$.

Factors of −120		Sum of Factors
−1		
−2		
−3		
−4		
−5		
−6		
−8		
−10		

$6x^2 + 7x - 20 = 6x^2$ ______ ______ $- 20$

$= 2x(\qquad) + 5(\qquad)$

$= (2x + 5)(\qquad)$

21. Factor using the AC method.

$2x^2 + 9x + 4$

Factors of _____		Sum of Factors

Multiply the factors to check your work.

22. Factor using the AC method.
$2x^2 - 11x + 14$

Factors of ____ **Sum of Factors**

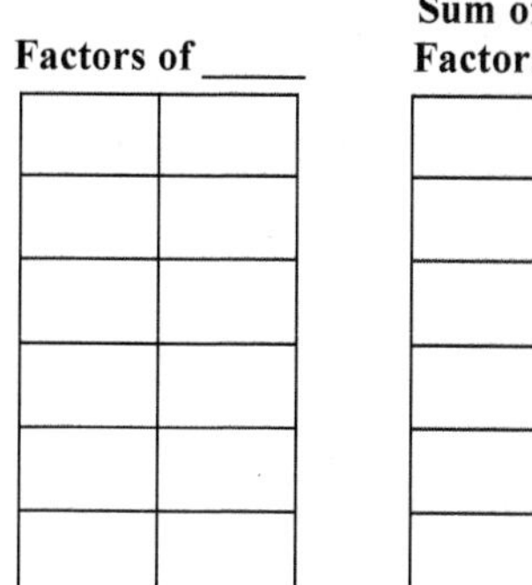

Multiply the factors to check your work.

23. Factor using the AC method.
$4x^2 - 17x - 15$

Factors of ____ **Sum of Factors**

Multiply the factors to check your work.

24. Factor using the AC method.
$10x^2 + 11x - 6$

Factors of ____ **Sum of Factors**

Multiply the factors to check your work.

Factor each trinomial by inspection or by the AC method. If it is prime, write "Prime" and justify your result.

25. $5x^2 + 29x + 20$

26. $4x^2 - 19x + 12$

27. $6y^2 - y - 35$

28. $6x^2 - 13x - 5$

29. $10x^2 - 19x + 6$

30. $8x^2 + 35x + 12$

31. $8x^2 - 26xy - 45y^2$

32. $12x^2 + 7xy - 10y^2$

First factor out -1.

33. $-2x^2 + 5x + 12$

34. $-3x^2 + 7x + 6$

First factor out the GCF.

35. $4x^2 + 20x - 56$

36. $20x^2 - 70x - 40$

37. $10x^3 + 25x^2 - 15x$

38. $6x^3 - 57x^2 + 105x$

39. $36x^3y^2 - 66x^2y^3 - 80xy^4$

40. $-24a^3b + 102a^2b^2 - 45ab^3$

METHODS FOR FACTORING $ax^2 + bx + c$	**ADVANTAGES**	**DISADVANTAGES**
Graphs	Visually displays the x-intercepts that correspond to the factors of the trinomial. If there are no x-intercepts, this indicates the polynomial is prime over the integers.	Can be time-consuming to select the appropriate viewing window and to approximate the x-intercepts. Because the x-intercepts are approximated, these factors should be checked.
Tables	Can see the zeros of the trinomial and can observe numerical patterns that are important in many applications. Spreadsheets allow us to use the power of computers to exploit this method.	Often requires insight or some trial and error in order to select the most appropriate table.
AC Method	This is a precise step-by-step process that can factor any trinomial of the form $ax^2 + bx + c$ or can identify the trinomial as prime. This method has the same steps used to multiply binomial factors but the steps are reversed.	Many trinomials with small integer coefficients can be factored by inspection and it is not necessary to write the table and all the steps of this method.
Trial-and-Error Method	Takes advantage of patterns and insights to quickly factor trinomials with small integer coefficients. Observing the mathematical patterns in these trinomials can improve your foundation for other topics.	For novices this process can be very challenging. It is important to examine all possibilities before deciding the trinomial is prime.

6.4 Lecture Guide: Factoring Special Forms

Objective 1: Factor perfect square trinomials.

Perfect Square Trinomials

Algebraically	Verbally	Algebraic Example
Square of a Sum $A^2+2AB+B^2=(A+B)^2$ **Square of a Difference** $A^2-2AB+B^2=(A-B)^2$	A trinomial that is the square of a binomial has: **1.** a first term that is a ___________ of the first term of the binomial. **2.** a middle term that is ___________ the product of the two terms of the binomial. **3.** a last term that is a ___________ of the last term of the binomial.	$x^2+12x+36=(x+6)^2$ $x^2-12x+36=(x-6)^2$

Complete the following table.

Products	Factors
1. $x^2+10x+25$	
2. $x^2-10x+25$	
3. $x^2-18x+81$	
4. $9x^2-42x+49$	
5. $9x^2+48xy+64y^2$	
6. $81x^2-18xy+y^2$	
7. $v^2-14v+49$	
8. $a^2+12ab+36b^2$	
9. $4m^2+12mn+9n^2$	

Determine the missing terms so each given trinomial is a perfect square trinomial.

10. x^2+ ________ $+9$

11. x^2+6x+ ________

12. x^2-16x+ ________

13. x^2- ________ $+64$

Objective 2: Factor the difference of two squares.

Difference of Two Squares

Algebraically	**Verbally**	**Algebraic Example**
$A^2 - B^2 = (A+B)(A-B)$	The difference of the squares of two terms factors as the ___________ of these terms times their ___________.	$x^2 - 49 = (x+7)(x-7)$

Factor each difference of two squares.

Products	**Factors**
14. $x^2 - 25$	
15. $x^2 - 64$	
16. $x^2 - 81$	
17. $16x^2 - 25$	
18. $9x^2 - 49y^4$	
19. $81x^2 - y^2$	
20. $a^2 - 36$	
21. $49m^2 - 100n^2$	
22. $9x^4 - 25$	

A common error is to factor an expression like x^2+9 as $x^2+9=(x+3)(x-3)$. Second degree polynomials that are the sum of two squares do not factor!

Sum of Two Squares

Algebraically	**Verbally**	**Algebraic Example**
$A^2 + B^2$ is prime*	The ___________ of two squares is prime.	$16x^2 + 9$ is prime.

* For binomials of degree greater than two this statement is not always true.

Factor each binomial if possible.

23. $x^2 + 81$

24. $x^2 - 81$

25. $4x^2 - 49$

26. $4x^2 + 49$

Objective 3: Factor the sum or difference of two cubes.

Find each product.

27. $(x+3)(x+3)(x+3)$

28. $(x+3)(x^2-3x+9)$

29. Which of the above is the correct factorization of x^3+27 ?

Sum or Difference of Two Cubes

Algebraically	Verbally	Algebraic Example
A^3+B^3 $=(A+B)(A^2-AB+A^2)$ A^3-B^3 $=(A-B)(A^2+AB+B^2)$	Write the binomial factor as the sum (or difference) of the two cube roots. Then use the binomial factor to obtain each term of the trinomial : **1.** The ____________ of the first term of the binomial is the first term of the trinomial factor. **2.** The opposite of the ____________ of the two terms of the binomial is the second term of the trinomial factor. **3.** The ____________ of the last term of the binomial is the third term of the trinomial factor.	x^3+8 $=(x+2)(x^2-2x+4)$ x^3-8 $=(x-2)(x^2+2x+4)$

Factor each sum or difference of two cubes.

Products	Factors
30. $x^3 - 27$	
31. $x^3 + 8$	
32. $x^3 - 125$	
33. $64x^3 + 125$	
34. $27x^3 - 1000$	
35. $8x^3 + 27y^3$	
36. $64x^3 - 27y^3$	

Factor each polynomial if possible. Don't forget to check for a GCF first.

37. $x^2 - 64$

38. $x^3 - 64$

39. $16x^2 - 49y^2$

40. $9x^2 - 36y^2$

41. $x^2 + 25$

42. $8x^3 + 125y^3$

43. $x^2+12x+36$

44. $x^2+13x+36$

45. $16x^2-56xy+49y^2$

46. $16x^2-70xy+49y^2$

47. $4x^2+20x+25$

48. $3x^2-75$

49. $4x^2-24xy+36y^2$

50. $-2x^2+20x-50$

51. $x^4 - 81$

52. $(3a+b)x^2 - (3a+b)y^2$

53. $x^2y^2 - 49z^2$

54. $-10x^2 - 60x - 90$

55. $(2a-1)x^2 - 4(2a-1)x - 21(2a-1)$

56. $(3a+b)^2 - (2a-5b)^2$

6.5 Lecture Guide:
Factoring by Grouping and a General Strategy for Factoring Polynomials

Objective 1: Factor polynomials by grouping.

Your goal upon finishing this section is to be able to use the various methods covered in this chapter to factor each given polynomial. In this first set of problems it is necessary to group pairs of terms as was done in Section 6.1. It may be necessary to reorder the terms to find a useful grouping. Completely factor the following polynomials.

1. $3x(a-b)-2(a-b)$

2. x^3+x^2-4x-4

3. $6xy+4x-3y-4$

4. $2x^3-18x-5x^2y+45y$

Factor each polynomial by using the special pattern $A^2-B^2=(A+B)(A-B)$.

5. $(3x-1)^2-25y^2$

6. $49x^2-(2a-b)^2$

Sometimes it is necessary to group 3 terms together. Start by trying to pick out a perfect square trinomial.

7. $36x^2+12x+1-49y^2$

8. $25x^2-y^2+6y-9$

Objective 2: Determine the most appropriate method for factoring a polynomial.

Strategy for Factoring a Polynomial Over the Integers

After factoring out the GCF (greatest common factor), proceed as follows.

Binomials: Factor special forms:

$A^2 - B^2 = (A+B)(A-B)$	____________ of Two Squares
$A^3 - B^3 = (A-B)(A^2 + AB + B^2)$	Difference of Two ____________
$A^3 + B^3 = (A+B)(A^2 - AB + A^2)$	____________ of Two Cubes
$A^2 + B^2$ is prime	The sum of two squares is____________ if A^2 and B^2 are only second-degree terms and have no common factor other than 1.

Trinomials: Factor the forms that are perfect squares:

$A^2 + 2AB + B^2 = (A+B)^2$	Perfect Square Trinomial; a Square of a Sum
$A^2 - 2AB + B^2 = (A-B)^2$	Perfect Square Trinomial; a Square of a Difference

Factor trinomials that are not perfect squares by inspection if possible; otherwise, use the trial-and-error method or the AC method.

Polynomials of Four or More Terms:
Factor by grouping

Factor each polynomial completely. If it is prime write "Prime" and justify your answer.

9. $25x^2 + 121y^2$

10. $3x^4 - 75x^2y^2$

11. $2x^3 - 18x^2 + 36x$

12. $x^4 - 10x^2 + 9$

13. $9x^2 - 30xy + 25y^2$

14. $25x^2 + 100x + 36$

15. $1250x^3 - 80y^3$

16. $-6x^5y - 48x^2y^4$

17. $x^2 - y^2 - 10x - 10y$

18. $60ax^2y - 12axy + 40x^2y - 8xy$

19. $x^3 - y^3 - 4x + 4y$

20. $x^3 - 8y^3 + 5x - 10y$

6.6 Lecture Guide: Solving Equations by Factoring

Objective 1: Identify a quadratic equation and write it in standard form.

Quadratic Equation

Algebraically	**Verbally**	**Algebraic Example**
If a, b, and c are real constants and $a \neq 0$, then $ax^2 + bx + c = 0$ is the ____________ form of a quadratic equation in x.	A quadratic equation in x is a ____________ -degree equation in x.	$3x^2 - 4x + 1 = 0$ is a quadratic equation with: Quadratic term : $3x^2$ Linear term : $-4x$ Constant term : 1

1. Determine which of these are quadratic equations in one variable.

(a) $x^3 + 2x^2 + x = 0$

(b) $x^2 - x - 20$

(c) $3 + 11x = -6x^2$

(d) $2x - 10 = 3x + 12$

2. Write each quadratic equation in standard form and identify a, b and c.

(a) $2x^2 - 3x^2 + 5 = 0$

Standard form: ____________________________

$a =$ ______, $b =$ ______, and $c =$ ______

(b) $3x^2 = 10$

Standard form: ____________________________

$a =$ ______, $b =$ ______, and $c =$ ______

(c) $8x^2 = 7x$

Standard form: ____________________________

$a =$ ______, $b =$ ______, and $c =$ ______

(d) $(x+3)(x-5) = 2$

Standard form: ____________________________

$a =$ ______, $b =$ ______, and $c =$ ______

Objective 2: Use factoring to solve selected second- and third-degree equations.

Zero-factor Principle

Algebraically	**Verbally**	**Algebraic Example**
If a and b are real numbers: If $a=0$ or $b=0$, then $ab=0$.	The product of zero and any other factor is ____________.	$0 \cdot a = 0$
If $ab=0$, then $a=0$ or $b=0$.	If the product of two factors is ____________, at least one of the factors must be zero.	If $(x-3)(x+4)=0$ then either $x-3=0$ or $x+4=0$.

Solve each equation.

3. $(x-9)(x+3)=0$

4. $(5n+2)(n-6)=0$

Solving Quadratic Equations by Factoring

Verbally	**Algebraic Example**
Step 1. Write the equation in ____________ form, with the right side zero.	$x^2+x=6$ $x^2+x-6=0$
Step 2. ____________ the left side of the equation.	$(x+3)(x-2)=0$
Step 3. Set each factor equal to _________.	$x+3=0$ or $x-2=0$
Step 4. Solve the resulting first-degree equations.	$x=-3$ or $x=2$

Solve each equation.

5. $x^2-10x+25=0$

6. $6x^2=-11x-3$

Solve each equation.

7. $25x^2 - 169 = 0$

8. $8x^2 = 3x$

9. $5x^2 + 46x = -9$

10. $8x^2 - 48x = 0$

11. $\frac{x^2}{12} - \frac{x}{3} - 1 = 0$

12. $x^3 = 2x^2 + 15x$

Solve each equation.

13. $(x-2)(x+7)=52$

14. $(2x-1)(x+3)=15$

15. $2x^2+4x-110=-2x-2$

16. $2x^2+6x=(3x+1)(x+3)$

Objective 3: Construct a quadratic equation with given solutions.

Equivalent Statements About Linear Factors of a Polynomial

For a real constant c and a real polynomial $P(x)$, the following statements are equivalent.

Algebraically	**Numerically**	**Graphically**
$x-c$ is a factor of $P(x)$	$P(c)=0$, that is c is a zero of $P(x)$	$(c,0)$ is an x-intercept of the graph of $y=P(x)$
$x=c$ is a solution of $P(x)=0$		

17. Solve $x^2-x-12=0$

18. Construct a quadratic equation in x with the given solutions.
$x=-3$ and $x=4$

Construct a quadratic equation in x with the given solutions.

19. $x=3$ and $x=-8$

20. $x=5$ and $x=-4$

21. $x=\frac{1}{3}$ and $x=-\frac{3}{4}$

22. $x=-\frac{2}{3}$ and $x=\frac{3}{5}$

Objective 4: Solve a quadratic inequality by using the x-intercepts of the corresponding graph.

Solutions of Equations and Inequalities

If $P(x)$ is a real polynomial and c is a real number, then:

Verbally	Algebraically	Graphically
c is a solution of $P(x)=0$	$P(c)=0$	$(c,0)$ is an x-intercept of the graph of $y=P(x)$.
c is a solution of $P(x)<0$	$P(c)<0$	At $x=c$, the graph is below the x-axis.
c is a solution of $P(x)>0$	$P(c)>0$	At $x=c$, the graph is above the x-axis.

23. Use the graph to help determine the solution of each equation and inequality.

(a) $x^2-7x-18=0$

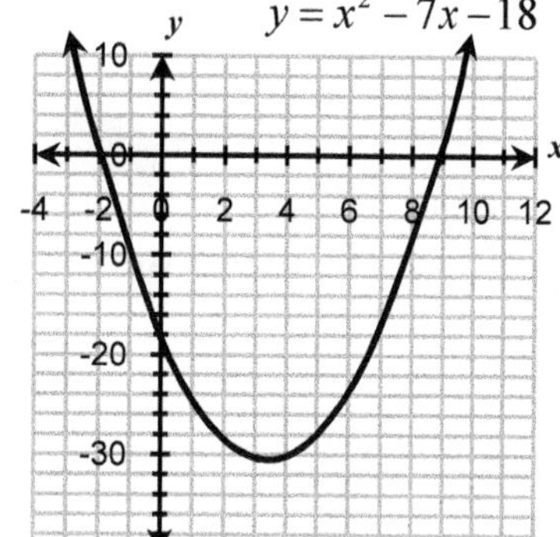

(b) $x^2-7x-18\le 0$

(c) $x^2-7x-18>0$

24. Use the graph to help determine the solution of each equation and inequality.

(a) $-x^2+4x+12=0$

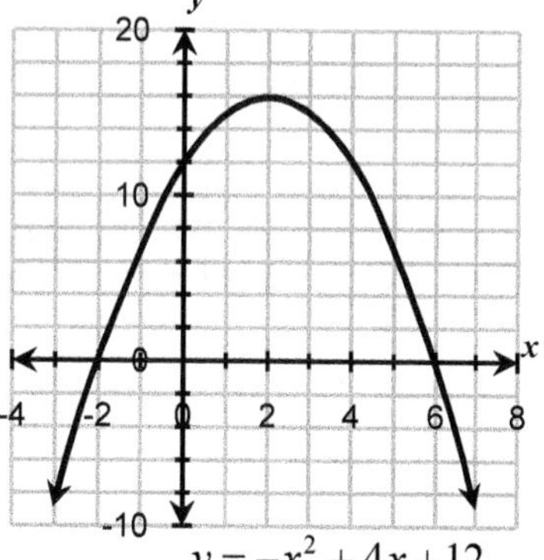

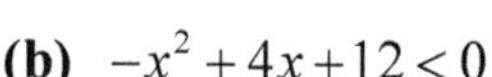

(b) $-x^2+4x+12<0$

(c) $-x^2+4x+12>0$

25. Use the table for $y = x^2 - 3x - 54$ to:

x	$y = x^2 - 3x - 54$
−6	0
−3	−36
0	−54
3	−54
6	−36
9	0
12	54

(a) Determine the zeros of $x^2 - 3x - 54$.

(b) Determine the x-intercepts of the graph of $y = x^2 - 3x - 54$.

(c) Factor $x^2 - 3x - 54$.

(d) Give the solutions of $x^2 - 3x - 54 = 0$.

26. Solve vs Simplify

In part **(a)**, solve the equation. In part **(b)**, perform the indicated operations and simplify the result.

(a) $-3(2x-1)(3x+4) = 0$ **(b)** $-3(2x-1)(3x+4)$

27. The length of the rectangle shown in the figure is 5 cm less than twice the width. Find the dimensions of this rectangle if the area is 63 cm^2.

w

Lecture Guides for

Chapter 7

Solving Quadratic Equations

7.1 Lecture Guide: Extraction of Roots and Properties of Square Roots

Objective 1: Solve quadratic equations by extraction of roots.

A **quadratic equation** is an equation that can be written in the ______________________ form $ax^2 + bx + c = 0$. The quadratic equation $x^2 - 3 = 0$ can be rewritten as $x^2 = 3$. The two solutions of this equation are the two numbers whose square is 3. We can denote these solutions by $x = \pm\sqrt{3}$. One solution is $x = \sqrt{3}$ and the other is $x = -\sqrt{3}$. Then notation "$\pm$" is read **plus or minus**. This process of solving quadratic equations is called **extraction of roots**.

Extraction of Roots	Example
If k is a positive real number, then the equation $x^2 = k$ has two real solutions, $x = \pm\sqrt{k}$: $x = \sqrt{k}$ and $x = -\sqrt{k}$.	$x^2 = 3$ $x = \sqrt{3}$ or $x = -\sqrt{3}$

Use extraction of roots to determine the exact solutions of each quadratic equation.

1. $x^2 = 16$

2. $x^2 = 7$

3. $2x^2 = 18$

4. $4x^2 - 12 = 0$

Objective 2: Use the product rule for radicals to simplify square roots.

Product Rule for Square Roots	**Algebraic Example**	**Numerical Example**
If $\sqrt{x}$ and $\sqrt{y}$ are both real numbers, then $\sqrt{xy} = \sqrt{x}\sqrt{y}$.	$\sqrt{12} = \sqrt{4 \cdot 3}$ $= \sqrt{4}\sqrt{3}$ $= 2\sqrt{3}$	√(12) 3.464101615 2√(3) 3.464101615

Simplify each square root using the product rule for square roots.

5. $\sqrt{4 \cdot 3}$

6. $\sqrt{40}$

7. $\sqrt{75}$

8. $\sqrt{32}$

Solve each quadratic equation using extraction of roots to obtain exact solutions. Then approximate any irrational solutions to the nearest hundredth.

9. $x^2 = 24$

10. $x^2 = 54$

11. $2x^2 = 64$

12. $5x^2 - 100 = 0$

Objective 3: Use the Quotient rule for radicals to simplify square roots.

Quotient Rule for Square Roots	Algebraic Example	Numerical Example
If $\sqrt{x}$ and $\sqrt{y}$ are both real numbers and $y>0$, then $\sqrt{\frac{x}{y}}=\frac{\sqrt{x}}{\sqrt{y}}$.	$\sqrt{\frac{25}{36}}=\frac{\sqrt{25}}{\sqrt{36}}$ $=\frac{5}{6}$	√(25/36)▸Frac 5/6 √(25)/√(36)▸Frac 5/6

Simplify each square root using the quotient rule for square roots.

13. $\sqrt{\frac{4}{25}}$

14. $\sqrt{\frac{11}{16}}$

15. $\frac{\sqrt{75}}{\sqrt{12}}$

16. $\frac{\sqrt{45}}{\sqrt{20}}$

Solve each quadratic equation using extraction of roots to obtain exact solutions. Then approximate any irrational solutions to the nearest hundredth.

17. $25x^2=11$

18. $4x^2-3=0$

Objective 4: Simplify expressions of the form $\frac{\sqrt{a}}{\sqrt{b}}$ ***by rationalizing the denominator.***

The process of rewriting a radical expression so that the _______________ does not have any radicals in it is called **rationalizing the denominator**. This process uses the fact that for $x > 0$, $\sqrt{x} \cdot \sqrt{x} = \sqrt{x^2} = x$.

Simplify each expression by rationalizing the denominator.

19. $\frac{3}{\sqrt{2}}$

20. $\frac{5}{\sqrt{6}}$

21. $\sqrt{\frac{5}{7}}$

22. $\sqrt{\frac{12}{20}}$

Solve each quadratic equation using extraction of roots to obtain exact solutions. Then approximate any irrational solutions to the nearest hundredth.

23. $5x^2 = 9$

24. $10x^2 - 15 = 0$

In the following box we extend the method of extraction of roots to a quadratic equation that contains the square of a binomial on the left side of the equation.

Extraction of Roots	Example
For a positive real number k and nonzero real numbers a and b: If $(ax+b)^2 = k$, then $ax+b = \pm\sqrt{k}$	$(2x+1)^2 = 3$ $2x+1=\sqrt{3}$ or $2x+1=-\sqrt{3}$

Solve each quadratic equation using extraction of roots.

25. $(x-1)^2 = 4$

26. $(x+3)^2 = 25$

27. $(2x+1)^2 + 5 = 41$

28. $(3x-2)^2 - 6 = 10$

Solve each equation.

29. $(2x-5)^2 = 9$

30. $2x-5=9$

31. Interest Rate A certificate of deposit earns an annual rate of interest r that is compounded twice a year. An investment of $1,300 grows to 1,412.85 by the end of 1 year. Use the formula $A = P\left(1 + \frac{r}{2}\right)^2$ to approximate the annual interest rate to the nearest tenth of a percent.

32. Checking a Solution You can check solutions of any equation by substitution or by graphing.

(a) In problem 18 from earlier in this section you were asked to solve $4x^2 - 3 = 0$. The solutions of this equation are $x = \pm\frac{\sqrt{3}}{2}$. Check these solutions of this by substitution.

(b) Check those same solutions by graphing $Y_1 = 4x^2 - 3$ and using the trace feature.

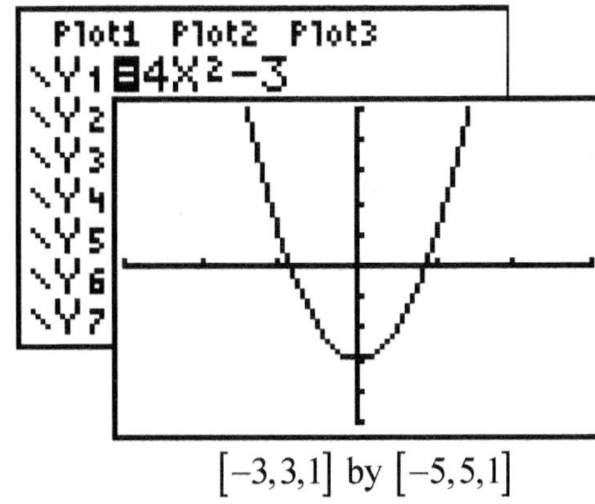

$[-3,3,1]$ by $[-5,5,1]$

7.2 Lecture Guide: Solving Quadratic Equations by Completing the Square

Objective 1: Determine the constant term in a perfect square trinomial.

A key to working the problems in this section is to recognize perfect square trinomials. Recall that:

Square of a sum: $A^2 + 2AB + B^2 = (\_\_\_\_\_\_\_\_\_)^2$

Square of a difference: $A^2 - 2AB + B^2 = (\_\_\_\_\_\_\_\_\_)^2$

Write each equation so the left side is expressed as the square of a binomial.

1. $x^2 + 6x + 9 = 5$

2. $x^2 - 8x + 16 = 81$

Fill in the missing constant term that is needed to make each expression a perfect square trinomial.

3. $x^2 + 10x +$ _____

4. $x^2 - 16x +$ _____

Objective 2: Solve quadratic equations by completing the square.

Rewrite the left side of each equation so that it is a perfect square, and then solve this equation by using extraction of roots.

5. $x^2 + 12x = 0$

6. $x^2 - 10x = -9$

Completing the Square

Step 1. Write the equation with the _______________ term on the right side.	Example: $x^2 - 6x - 12 = 0$
Step 2. Divide both sides of the equation by the coefficient of x^2 to obtain a coefficient of ______ for x^2.	
Step 3. Take one-half of the coefficient of x, square this number, and add the result to _______________ sides of the equation.	
Step 4. Write the left side of the equation as a perfect _______________.	
Step 5. Solve this equation by extraction of _______________.	

Solve the following equations using the method of completing the square.

7. $x^2 + 2x + 4 = 7$

8. $x^2 - 8x + 5 = 0$

9. $3x^2 + 9x = 3$

10. $2x^2 - 20x - 4 = 1$

11. $x^2 - \frac{7}{2}x = \frac{5}{2}$

12. $x(14 - x) = 40$

13. Use the graph of $y = x^2 - 5$ to:

(a) Solve $x^2 - 5 = 0$.

(b) Solve $x^2 - 5 < 0$.

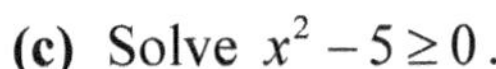

(c) Solve $x^2 - 5 \geq 0$.

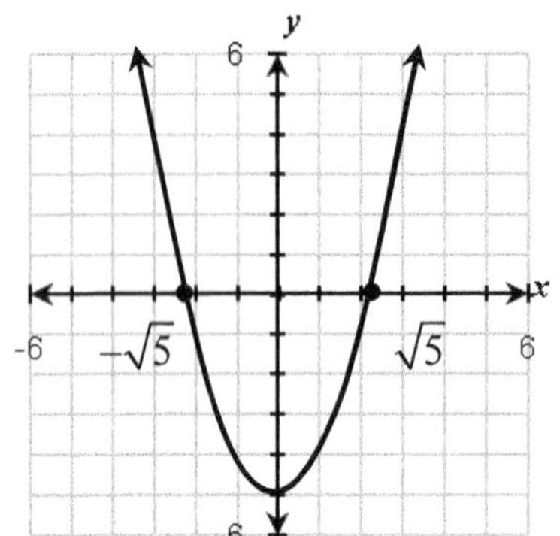

14. Construct a quadratic equation in x that has the given solutions.

(a) $x = \frac{3}{4}$ and $x = -\frac{3}{4}$

(b) $x = \frac{\sqrt{3}}{4}$ and $x = -\frac{\sqrt{3}}{4}$

15. A square picture is surrounded by a mat and then framed. The width of the square mat is 1.5 times the width of the picture. If the area covered by the mat is 145 in^2, determine the width of the picture. Round to the nearest tenth of an inch.

(a) Identify the variable:

Let w = the ________________ of the picture in inches

Let ____________ = the ________________ of the mat in inches

(b) Write the word equation:

Area covered by mat = Total area covered by mat and picture - ________________________

(c) Translate the word equation into an algebraic equation:

________________________ = ________________________ - ________________________

(d) Solve this equation:

(e) Write a sentence that answers the question:

(f) Is this answer reasonable?

7.3 Lecture Guide: Using the Quadratic Formula to Find Real Solutions

Objective 1: Use the quadratic formula to solve quadratic equations with real solutions.

Methods for solving quadratic equations we have covered so far:

- Tables --- the zeros in a table correspond to the solutions
- Graphs --- the x-intercepts correspond to the solutions
- Factoring --- the factors of the polynomial correspond to the solutions
- Extraction of roots
- Completing the square

The quadratic formula can be used to solve any quadratic equation --- an all purpose tool. The quadratic formula can be derived using completing the square as shown below.

Solving $ax^2+bx+c=0$ by Completing the Square

Step 1. Write the equation with the constant term on the right side.

$$ax^2+bx+c=0$$
$$ax^2+bx=$$

Step 2. Divide both sides of the equation by the coefficient of x^2 to obtain a coefficient of 1 for x^2.

$$\frac{ax^2}{a}+\frac{bx}{a}=$$

$$x^2+\frac{b}{a}x=$$

Step 3. Take one-half of the coefficient of x, square this number, and add the result to both sides of the equation.

$$x^2+\frac{b}{a}x+\qquad\qquad=$$

Step 4. Write the left side of the equation as a perfect square.

$$\left(x+\frac{b}{2a}\right)^2=$$

Step 5. Solve this equation by extraction of roots.

$$x+\frac{b}{2a}=\pm\sqrt{\frac{\qquad}{\qquad}}$$

$$x=-\frac{b}{2a}\pm\frac{\sqrt{b^2-4ac}}{2a}$$

$$x=\frac{-b\pm\sqrt{b^2-4ac}}{2a}$$

The **Quadratic Formula**, which gives the solutions of the quadratic equation $ax^2 + bx + c = 0$ with real coefficients *a*, *b*, and *c*, when $a \neq 0$, is:

Use the quadratic formula to determine the exact solutions of each quadratic equation. Then approximate each solution to the nearest hundredth.

1. $x^2 + 2x - 2 = 0$

2. $4x^2 + 6x + 1 = 0$

3. $2x^2 + 3x = 3$

4. $-x^2 + 3x = -1$

Objective 2: Use the discriminant to determine the nature of the solutions of a quadratic equation.

A part of the quadratic formula that determines the nature of the solutions is the expression under the radical symbol. $b^2 - 4ac$ is called the **discriminant**. This expression is important since $\sqrt{b^2 - 4ac}$ is not a real number if $b^2 - 4ac$ is ________________________.

The Nature of the Solutions of a Quadratic Equation

There are three possibilities for the solutions of $ax^2 + bx + c = 0$.

Value of the Discriminant	Solutions of $ax^2 + bx + c = 0$	The Parabola $y = ax^2 + bx + c$	Graphical Example
1. $b^2 - 4ac > 0$	Two distinct real solutions	Two x-intercepts	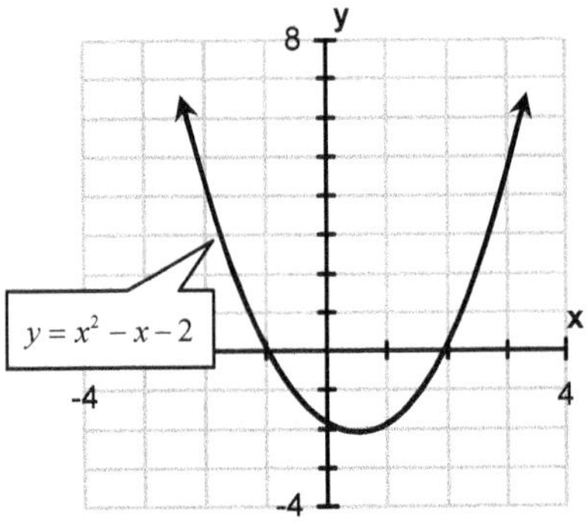 $y = x^2 - x - 2$
2. $b^2 - 4ac = 0$	A double real solution	One x-intercept with the vertex on the x-axis	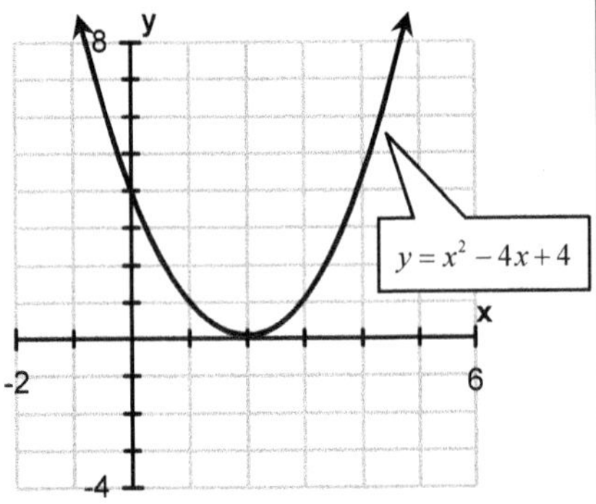 $y = x^2 - 4x + 4$
3. $b^2 - 4ac < 0$	Neither solution is real; both solutions are complex numbers with imaginary parts. These solutions will be complex conjugates.*	No x-intercepts	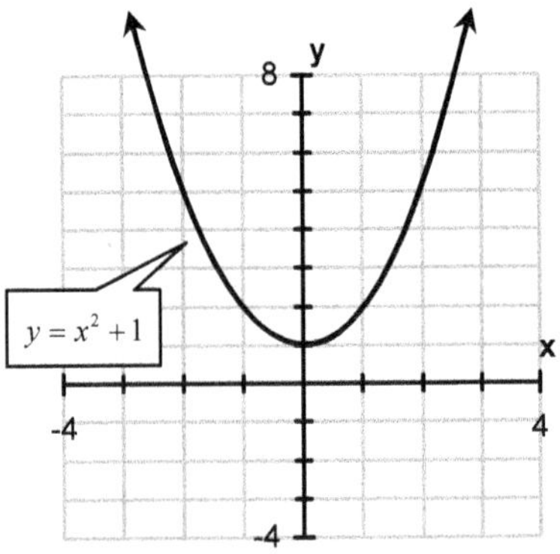 $y = x^2 + 1$

* Complex numbers are covered in Section 7.5.

Compute the value of each discriminant, $b^2 - 4ac$, and determine the nature of the solutions.

5. Equation $3x^2 + 5x - 2 = 0$

Graph

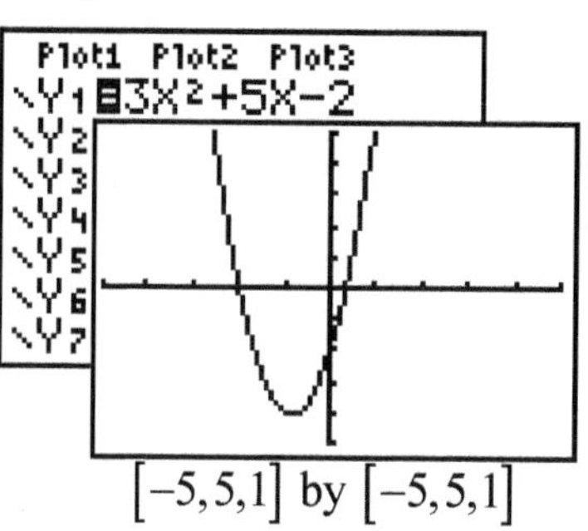

$[-5,5,1]$ by $[-5,5,1]$

Discriminant

Nature of Solutions

6. Equation $4x^2 - 12x + 9 = 0$

Discriminant

Nature of Solutions

Graph

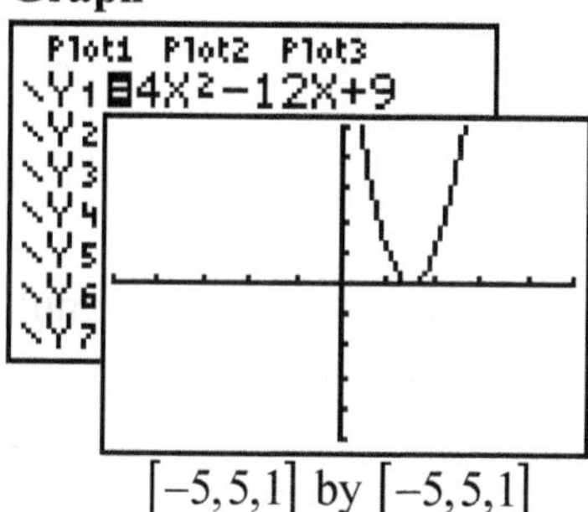

$[-5,5,1]$ by $[-5,5,1]$

7. Equation $x^2 - 6x + 10 = 0$

Discriminant

Nature of Solutions

Graph

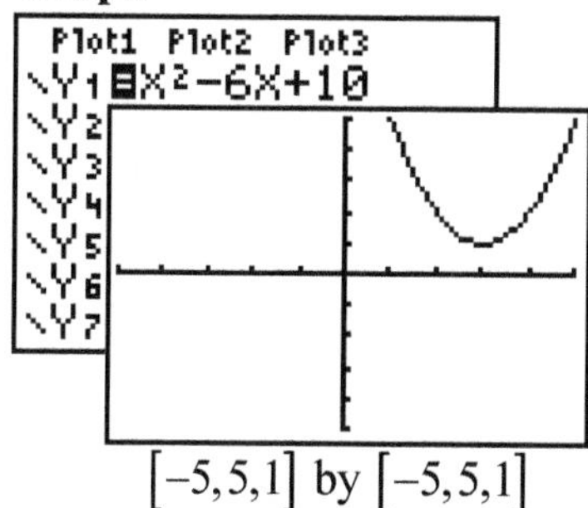

$[-5,5,1]$ by $[-5,5,1]$

8. Use the quadratic formula to determine the exact solutions of the quadratic equation $x^2 - x - 4 = 0$. Then approximate each solution to the nearest hundredth.

9. Use the graph and the solution from problem 8 to solve the following inequalities.

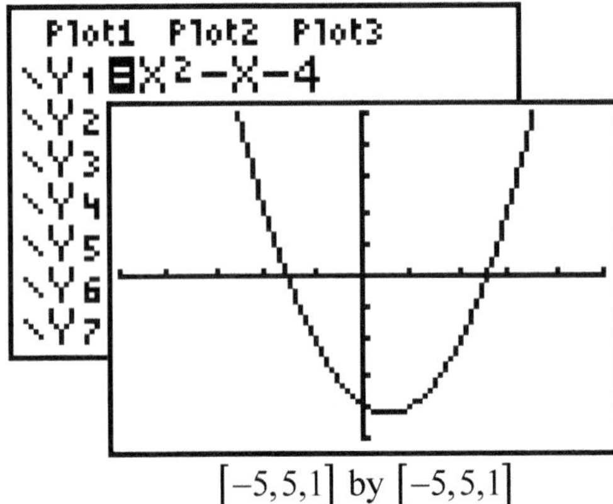

[−5,5,1] by [−5,5,1]

(a) $x^2 - x - 4 \leq 0$

(b) $x^2 - x - 4 \geq 0$

10. Use the quadratic formula to determine the exact solutions of the quadratic equation $-x^2 + 4x + 1 = 0$. Then approximate each solution to the nearest hundredth.

11. Use the graph and the solution from problem 10 to solve the following inequalities.

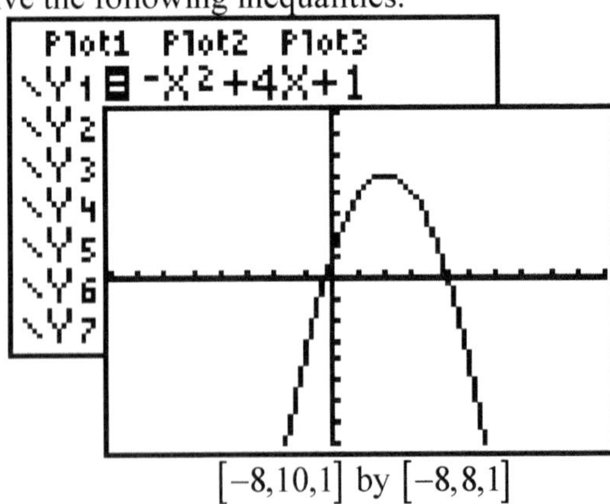

[−8,10,1] by [−8,8,1]

(a) $-x^2 + 4x + 1 < 0$

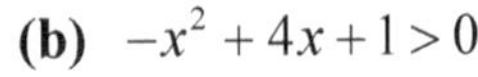

(b) $-x^2 + 4x + 1 > 0$

12. The weekly profit in dollars for selling x bottles of hand lotion is given by $P(x) = -2x^2 + 50x - 20$.

(a) Determine the overhead costs for the company. Hint: Evaluate $P(0)$.

(b) Determine the break-even values for the company. Hint: Determine to the nearest unit when $P(x) = 0$.

13. Construct a quadratic equation in x that has solutions of $x = \dfrac{4+\sqrt{5}}{2}$ and $x = \dfrac{4-\sqrt{5}}{2}$.

7.4 Lecture Guide: Applications of Quadratic Equations

Objective 1: Use quadratic equations to solve word problems.

Strategy for Solving Word Problems

Step 1. Read the problem carefully to determine what you are being asked to find.
Step 2. Select a ____________ to represent each unknown quantity. Specify precisely what each variable represents and note any restrictions on each variable.
Step 3. If necessary, make a ____________ and translate the problem into a word equation or a system of word equations. Then translate each word equation into an ____________ equation.
Step 4. Solve the equation or system of equations, and answer the question completely in the form of a sentence.
Step 5. Check the ________________ of your answer.

1. Find two consecutive odd integers whose product is 195.

(a) Identify the variable:

Let n = the smaller odd integer

Let ________________ = the larger odd integer

(b) Write the word equation:

The ________________ of the two integers is ________________ .

(c) Translate the word equation into an algebraic equation:

(d) Solve this equation:

(e) Write a sentence that answers the question:

(f) Is this answer reasonable?

2. The sum of the squares of two consecutive even integers is 244. Find these integers.

3. One number is two less than six times another number. Find these numbers if their product is 48.

4. **Area of a Rectangle** The length of a rectangle is 3 cm more than 2 times the width. Find the dimensions of this rectangle if the area is 65 cm^2.

w

l

(a) Identify the variable:

Let w = the width of the rectangle in __________

Let ________________ = the ____________ of the rectangle in ________

(b) Write the word equation:

The ________________ is __________________ .

(c) Translate the word equation into an algebraic equation:

(d) Solve this equation:

(e) Write a sentence that answers the question:

(f) Is this answer reasonable?

5. **Finding Dimensions** If each side of a square is increased by 10 inches, the total area of both the new square and the original square will be 210 square inches. What is the length of each side of the original square? Round to the nearest tenth of an inch.

6. **Height of a Baseball** The height in feet of a baseball is given by $h(t) = -16t^2 + 50t + 4$ where t represents the time in seconds the ball is in flight.
(a) Determine the number of seconds before the baseball hits the ground at a height level of 0 ft.

(b) Determine the time interval for which the height of the ball is above 20 ft.

7. **Calculating an Interest Rate** If the formula computing the amount A of an investment of principal P invested at interest rate r for 1 year and compounded semiannually is $A = P\left(1+\frac{r}{2}\right)^2$, what interest rate is necessary for \$12,000 to grow to \$12,484.80?

Objective 2: Use the Pythagorean Theorem.

The Pythagorean Theorem

If triangle ABC is a right triangle, then:

Algebraically	**Verbally**
$a^2 + b^2 = c^2$	The sum of the areas of the squares formed on the legs of a right triangle is equal to the area of the square formed on the hypotenuse.

(The converse of this theorem is also true. If $a^2 + b^2 = c^2$, then triangle ABC is a right triangle.)

Use the Pythagorean Theorem to find the length of the side that is not given. Round to the nearest hundredth.

8.

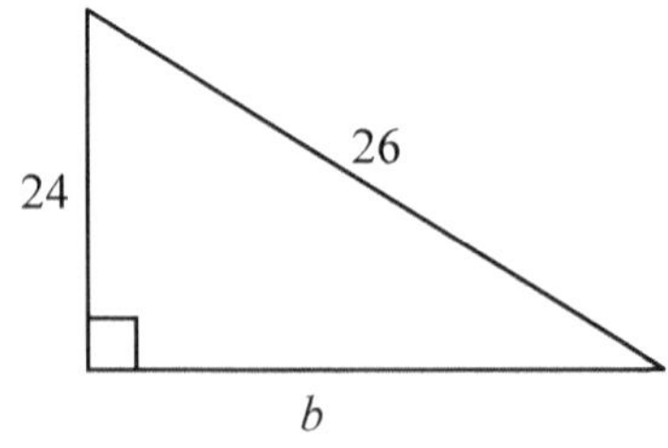

9.

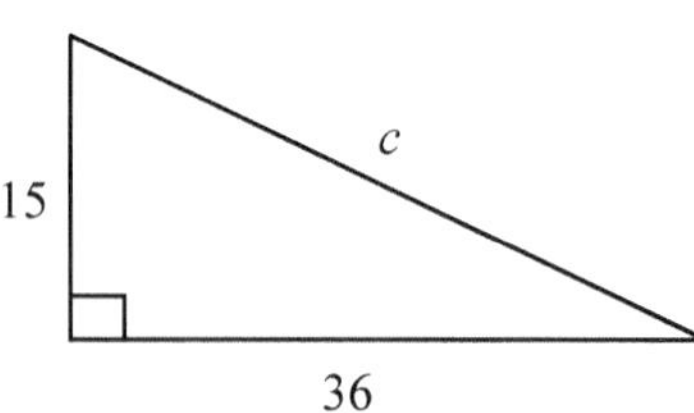

10.

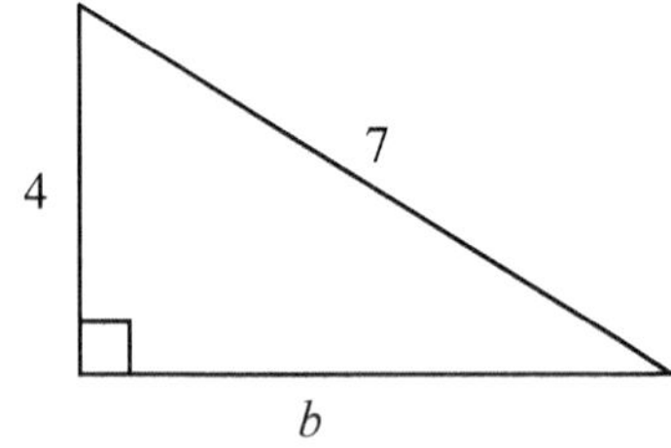

11.

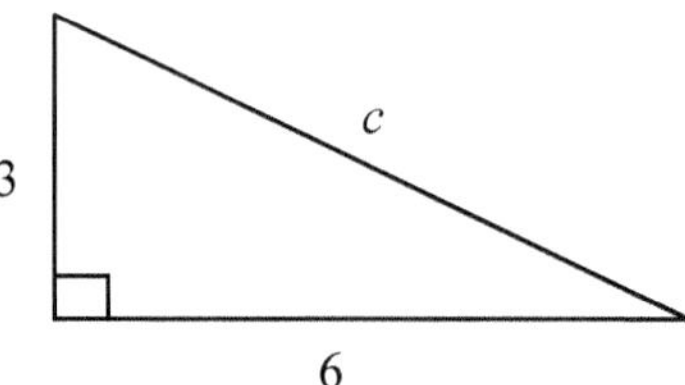

12. A 42-inch wide screen television is advertised as having a 8:5 aspect ratio. (The aspect ratio is the ratio of the width to the height of the screen.) Determine the actual dimensions and the viewable area of this television.

13. Two airplanes depart simultaneously from an airport. One flies due south; the other flies due east at a rate 20 miles per hour faster than that of the first airplane. After 4 hours radar indicates that the airplanes are 800 miles apart. What is the ground speed of each airplane? Round the speed to the nearest whole number.

7.5 Lecture Guide: Complex Numbers and Solving Quadratic Equations with Complex Solutions

Objective 1: Express complex numbers in standard form.

The Imaginary Number *i*

The imaginary number i is defined as $i = \sqrt{-1}$. So:

$i^2 =$ $\quad\quad$ $i^3 =$ $\quad\quad$ $i^4 =$

Notice that any power of i can be simplified to $i, -1, -i,$ or 1. Simplify each power of i.

1. i^5 $\quad\quad$ **2.** i^{17} $\quad\quad$ **3.** i^{22}

4. i^{60} $\quad\quad$ **5.** i^{801} $\quad\quad$ **6.** i^{107}

Complex Numbers

If a and b are real numbers and $i = \sqrt{-1}$, then:

Algebraic Form	**Numerical Example**
$a + bi$ is a complex number with a ____________ term a and an ____________ term bi.	$2 + 7i$ has a real term ______ and an imaginary term ______.

Simplify each expression and write the result in the standard $a + bi$ form.

7. $\sqrt{-36}$ $\quad\quad$ **8.** $\sqrt{-6}$ $\quad\quad$ **9.** $\sqrt{(-6)(-6)}$

10. $-\sqrt{-81}$ $\quad\quad$ **11.** $\sqrt{-81}\sqrt{-81}$ $\quad\quad$ **12.** $-\sqrt{81} - \sqrt{-81}$

13. $\dfrac{\sqrt{-100}}{\sqrt{25}}$ $\quad\quad$ **14.** $\sqrt{-6}\left(\sqrt{5} + \sqrt{-6}\right)$ $\quad\quad$ **15.** $-\sqrt{36} + \sqrt{-25} - \sqrt{49}$

Objective 2: Add, subtract, multiply, and divide complex numbers.

The arithmetic of complex numbers is very similar to the arithmetic of binomials.

Addition of Binomials

$$(5x+7y)+(3x-4y)=(5x+3x)+(7y-4y)$$
$$=8x+3y$$

Addition of Complex Numbers

$$(5+7i)+(3-41)=(5+3)+(7i-4i)$$
$$=8+3i$$

Simplify each expression and write the result in the standard $a+bi$ form.

16. $(3-7i)+(9+8i)$

17. $(4+7i)-(3-2i)$

18. $6(7-3i)$

19. $(3i)(10i)$

20. $2i(7-9i)$

21. $(4+3i)(5-i)$

22. $(7+6i)(3-8i)$

23. $(2-5i)^2$

Complex Conjugates: The conjugate of $a+bi$ is $a-bi$.

Write the conjugate of each expression. Then multiply the expression by its conjugate.

Expression	Conjugate	Product
24. $5-2i$		
25. $-8i$		

Dividing Complex Numbers The fact that the product of a complex number and its conjugate is always a real number plays a key role in the division of complex numbers as outlined in the following box.

Steps for Dividing Complex Numbers

Step 1. Write the division problem as a fraction. Example: $(3-i)\div(1+2i)$

Step 2. Multiply both the numerator and the denominator by the conjugate of the denominator.

Step 3. Simplify the result, and express it in standard $a+bi$ form.

Perform the indicated operations and express the result in $a+bi$ form.

26. $50\div(1+3i)$

27. $\dfrac{2}{3-4i}$

28. $\dfrac{5+3i}{2-7i}$

29. $\dfrac{6+8i}{1+i}$

30. Determine whether or not $x = 2 - 5i$ is a solution of the equation $x^2 - 6x + 13 = 0$.

31. Determine whether or not $x = 3 - 2i$ is a solution of the equation $x^2 - 6x + 13 = 0$.

Objective 3: Solve a quadratic equation with imaginary solutions

Recall solving quadratic equations by **extraction of roots** from Section 7.1: The solutions of $x^2 = k$ are $x = \sqrt{k}$ and $x = -\sqrt{k}$.

Solve each quadratic equation by extraction of roots.

32. $x^2 = -16$

33. $x^2 = -7$

34. $(x-1)^2 = -4$

35. $(x+3)^2 = 49$

Solve each quadratic equation by extraction of roots.

36. $(2x+3)^2 = -49$

37. $(3x-2)^2 = -10$

Recall solving quadratic equations by the **Quadratic Formula** from Section 7.3: The solutions of the quadratic equation $ax^2 + bx + c = 0$ with real coefficients *a*, *b*, and *c*, when $a \neq 0$, are

$$x = \frac{-b \pm \sqrt{b^2 - 4ac}}{2a}.$$

Use the quadratic formula to solve each equation.

38. $x^2 - 2x + 6 = 0$

39. $3x^2 - 2x = 4$

40. $x^2 - 10x = -25$

41. $x(3x-4) = -5$

42. $(x-2)(x^2+2x+4)=0$
(Hint: Use the zero factor principle.)

Construct a quadratic equation in x that has the given solutions.

43. $x=5i$ and $x=-5i$

44. $x=3+i$ and $x=3-i$

45. Determine the discriminant of each of these quadratic equations and then determine the solution of each equation.

Equation	**Discriminant**	**Solution**
(a) $x^2+1=0$		
(b) $x^2+2x+1=0$		
(c) $x^2-1=0$		

Lecture Guides for

Chapter 8

Functions: Linear, Absolute Value, and Quadratic

8.1 Lecture Guide: Functions and Representations of Functions

Objective 1: Identify a function and determine its domain and range.

Function

A **function** is a correspondence that matches each ____________ value with exactly ____________ value of the output variable. The set of all possible input values is called the ____________ of the function or the set of **independent** values. The set of all output values is called the ____________ of the function or the set of **dependent** values.

1. Classify each correspondence as either a function or as a correspondence that is not a function. For each function identify the domain and range. Circle the correct choice.

(a) D R

$5 \to 13$

$5 \to 20$

$7 \to 51$

Function? Yes / No

(b) D R

$1 \to 2$

$3 \to -7$

$8 \to -12$

Function? Yes / No

(c) D R

$-2 \to 9$

$1 \to 9$

$3 \to 3$

Function? Yes / No

2. Give the domain and range of the following function, and convert this function given in ordered pair notation to mapping notation and then graph this function.

Ordered-Pair Notation

$\{(3,4),\ (2,-1),\ (0,1),\ (5,4)\}$

Domain and Range

Mapping Notation

Graph

y, 6, -6, 6, x, -6

Domain and Range from a Graph

Domain: The domain of a function is the projection of its graph onto the _____-axis.

Range: The range of a function is the projection of its graph onto the _____-axis.

Give the domain and range of each graph:

3.

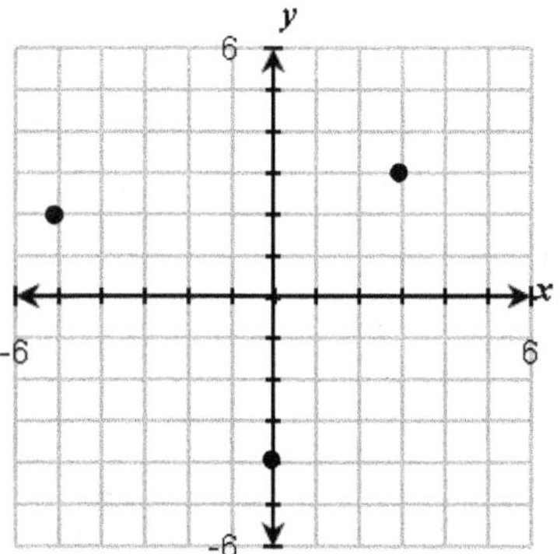

Domain:

Range:

4.

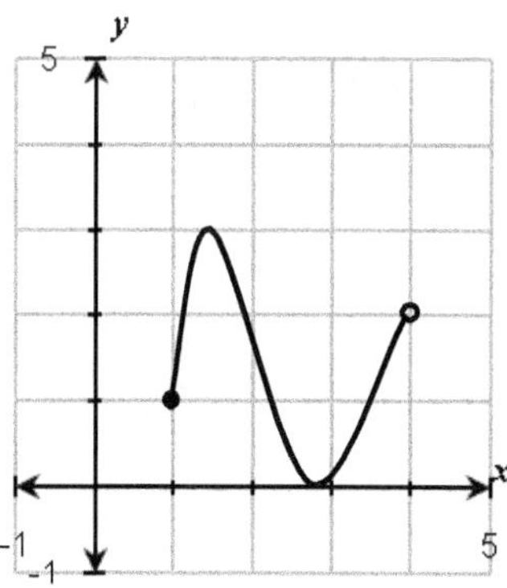

Domain:

Range:

5.

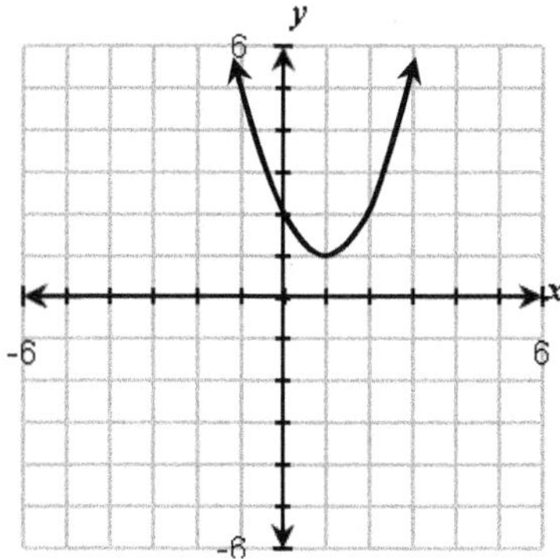

Domain:

Range:

6.

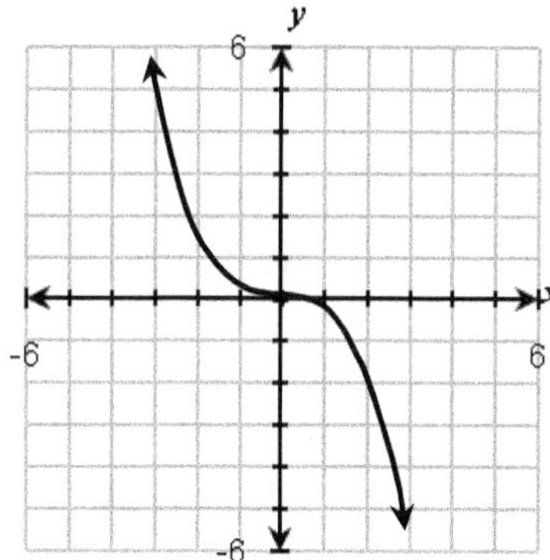

Domain:

Range:

Vertical Line Test

A graph represents a function if it is impossible to have any vertical line intersect the graph at more than _______________ point.

Use the vertical line test to determine whether each graph represents a function. Circle the correct choice.

7.

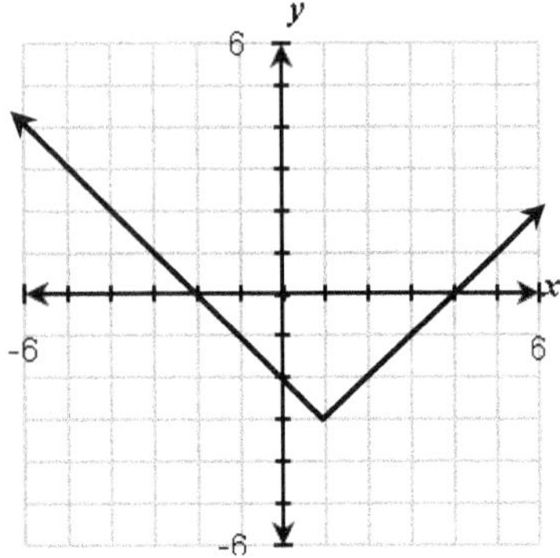

Function? Yes / No

8.

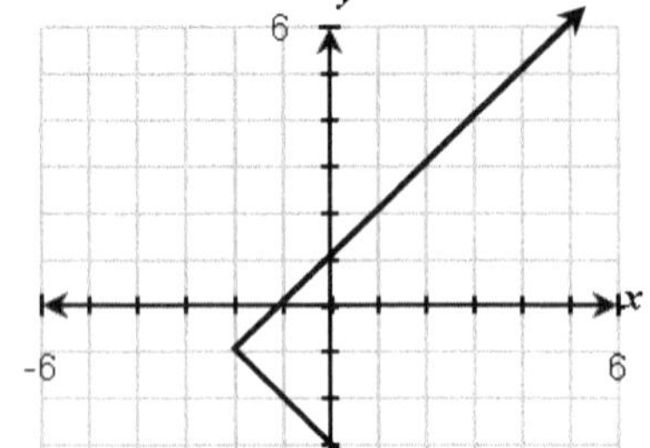

Function? Yes / No

Objective 2: Use function notation.

Function Notation

The expression $f(x)$ is read as "f of x" or "the value of f at x." The letter f names the function, and the variable x represents an ____________ value from the domain. The notation $f(x)$ represents a specific ____________ value corresponding to x.

Graph each linear function:

	Evaluate $f(x)$	**Ordered Pairs**	**Graph of Line**
9. $f(x) = -2x + 3$	$f(0) =$		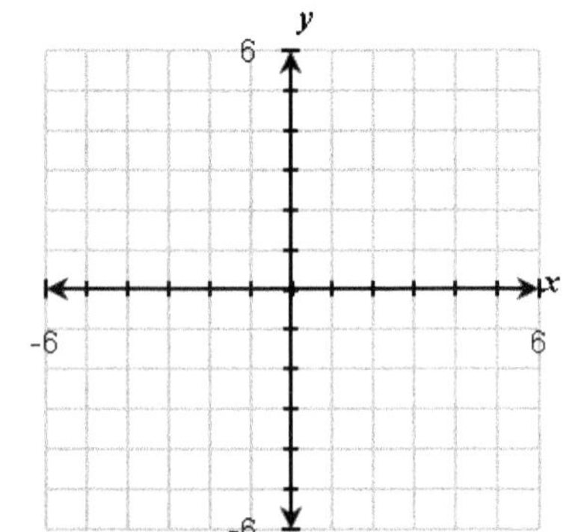
	$f(1) =$		
	$f(4) =$		

	Evaluate $f(x)$	**Ordered Pairs**	**Graph of Line**
10. $f(x) = -\frac{2}{5}x + 1$	$f(-5) =$		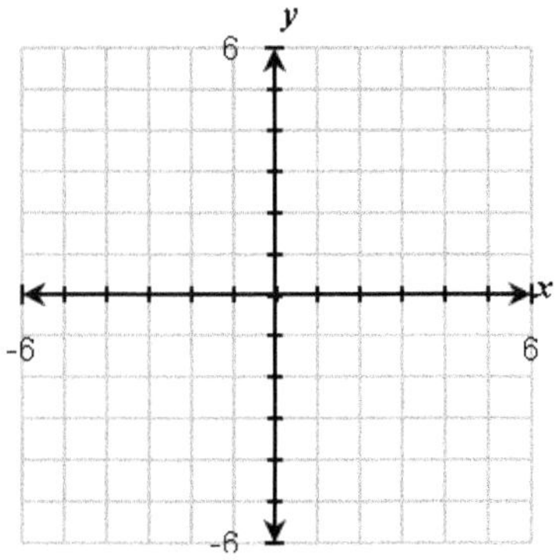
	$f(0) =$		
	$f(5) =$		

11. Note that these functions are in the form $y = mx + b$, which is called the ______________________ - ______________________ form of a line.

12. Use the table to answer the following questions:

x	y
–3	21
–1	5
1	–3
2	–4
6	12

(a) Evaluate $f(2)$.

(b) Find the value of x for which $f(x) = -3$.

(c) Evaluate $f(-1)$.

(d) Find the value of x for which $f(x) = 12$.

13. Use the graph to answer the following questions:

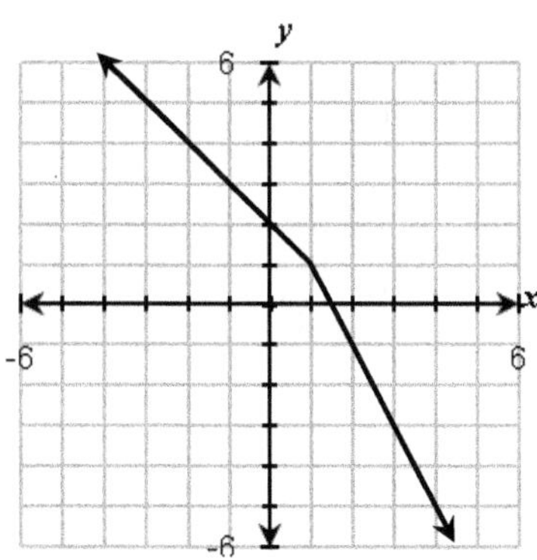

(a) Evaluate $f(-3)$.

(b) Find the value of x for which $f(x) = -3$.

(c) Evaluate $f(2)$.

(d) Find the value of x for which $f(x) = 2$.

8.2 Lecture Guide: Linear Functions

Key Characteristics of a Linear Function

1. The graph of a linear function is a ____________ line.

2. The equation is first degree and can be written in the slope-intercept form $f(x) = mx + b$ with slope ______ and y-intercept ______.

3. Linear functions have a constant rate of change. The slope is the same between any two ____________ on the line.

4. The domain of all linear functions is ______. The range of all linear functions is also $\mathbb{R}$ unless the function is a constant function, whose graph is a horizontal line.

Linear Function

Algebraically	**Numerical Example**	**Graphical Example**	**Verbally**
A function of the form $f(x) = mx + b$ is called a linear function.	$f(x) = 2x - 5$ x | $y = f(x)$ -1 | -7 0 | -5 1 | -3 2 | -1 3 | 1 4 | 3 5 | 5	$f(x) = 2x - 5$	In the table for $y = 2x - 5$, each 1-unit increase in x produces a 2-unit increase in y. The graph of $y = 2x - 5$ is a straight line. Each point on this line satisfies this equation. This line rises 2-units for each 1-unit move to the right.

Objective 1: Determine the slope of a line.

From Section 3.1, we define slope of a line through two points as the ratio of the change in y to the change in x. This rise over the run can be expressed algebraically by $m = \frac{y_2 - y_1}{x_2 - x_1}$ or $m = \frac{\Delta y}{\Delta x}$. A line defined by the slope-intercept form $f(x) = mx + b$ has a slope m.

Determine the slope of each line.

5. $f(x) = -\frac{3}{5}x + 7$

6. $f(x) = -3$

7. $f(x) = 4x - 5$

8. Calculate the slope of the line in the graph.

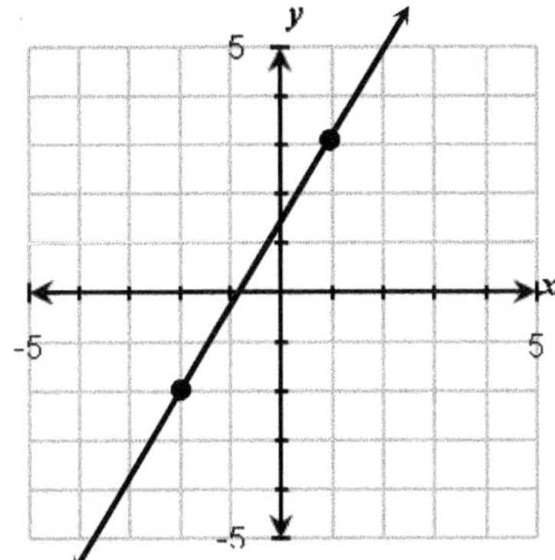

9. Calculate the slope of the line in the graph.

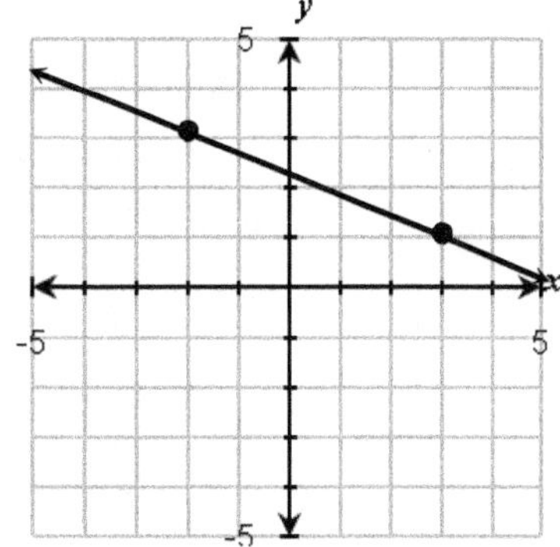

10. Calculate the slope of the line in the graph.

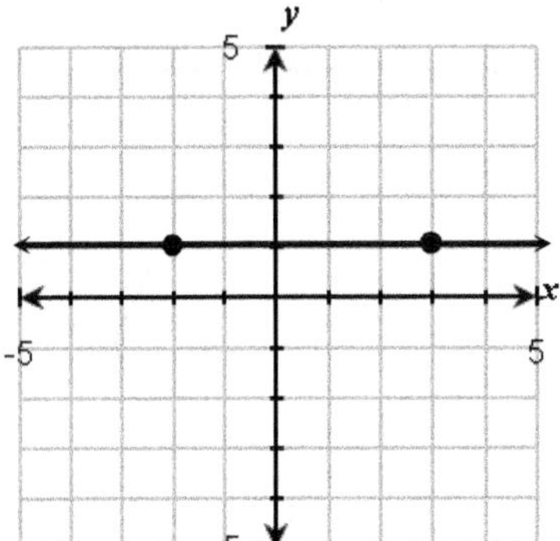

11. Calculate the slope of the line in the graph.

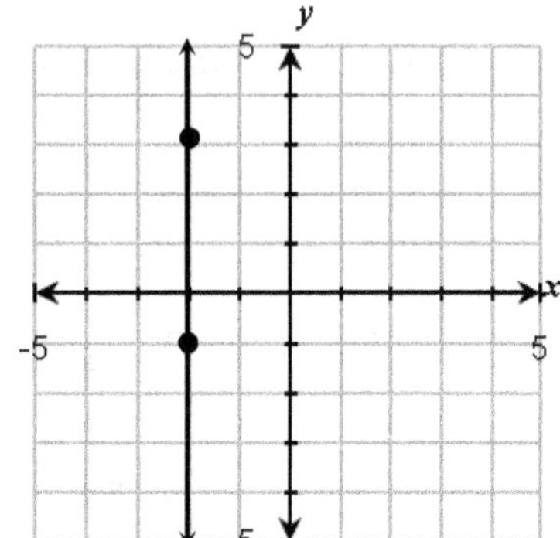

12. Calculate the slope of the line containing the points in the table.

x	y
0	2
5	5
10	8
15	11
20	14
25	17
30	20

13. Calculate the slope of the line containing the points in the table.

x	y
−3	4
0	2
3	0
6	−2
9	−4
12	−6
15	−8

Interpreting Slopes

Positive Slope

Algebraically

In $f(x) = mx + b$, m will be positive.

Example: $f(x) = \frac{1}{2}x - 1$

Graphically

The line will slope ____________ to the right.

Example:

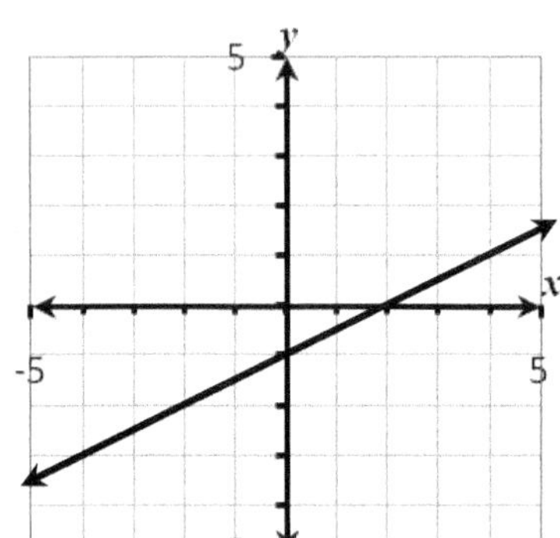

Numerically

The y-values will ____________ as the x-values increase.

Example:

x	y
-4	-3
-2	-2
0	-1
2	0
4	1

Zero Slope

Algebraically

In $f(x) = mx + b$, m will be 0 and $f(x) = b$.

Example: $f(x) = 4$

Graphically

This is a ____________ line that does not slope upward or downward.

Example:

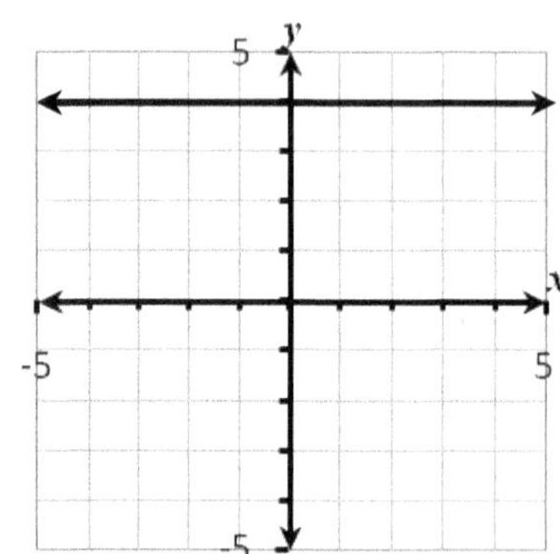

Numerically

The y-values will ________ ____________ as the x-values change.

Example:

x	y
-2	4
-1	4
0	4
1	4
2	4

Negative Slope

Algebraically

In $f(x) = mx + b$, m will be negative.

Example: $f(x) = -2x + 1$

Graphically

The line slopes ____________ to the right.

Example:

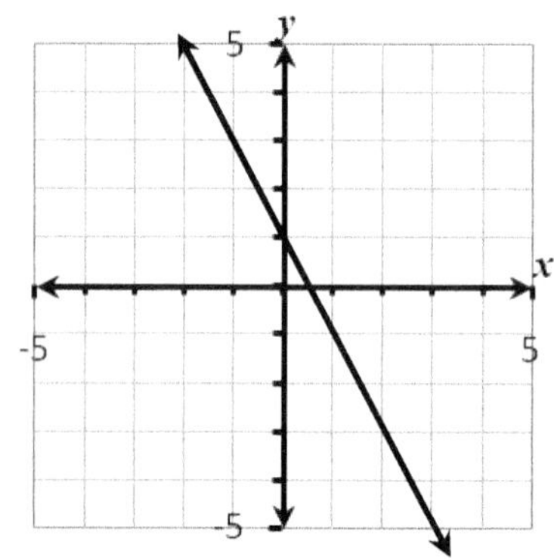

Numerically

The y-values will ____________ as the x-values increase.

Example:

x	y
-2	5
-1	3
0	1
1	-1
2	-3

The equation of a vertical line is of the form $x = k$. This equation does not represent a function --- it fails the vertical line test. The slope of a vertical line is ____________.

Objective 2: Sketch the graph of a linear function.

Use the slope and y-intercept to graph each line.

14. $f(x) = \frac{3}{4}x - 5$

Slope: ______

y-intercept: ______

Graph:

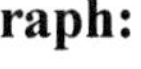

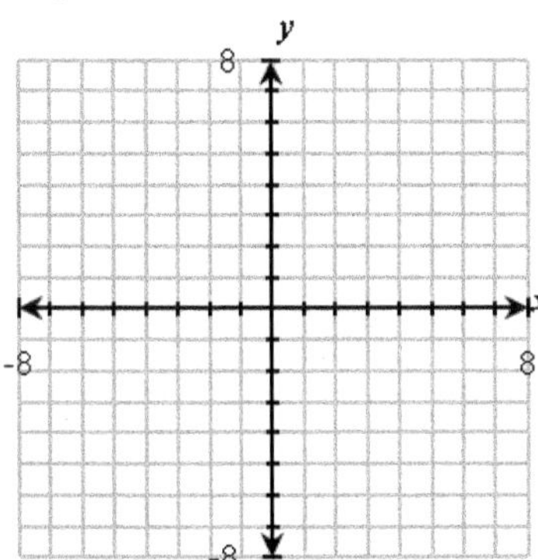

15. $f(x) = -\frac{1}{2}x + 3$

Slope: ______

y-intercept: ______

Graph:

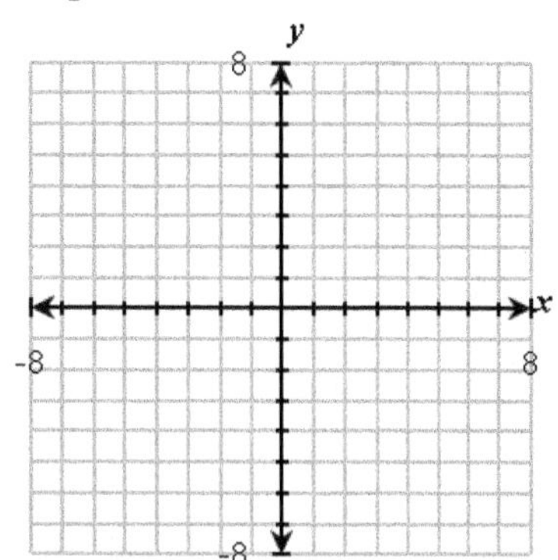

Objective 3: Write the equation of the line through given points.

Determine the equation of the line in the form $f(x) = mx + b$ that passes through each pair of points.

16. $(-3, 4)$ and $(0, 7)$

17. $(1, -2)$ and $(4, -6)$

All the points listed in the table lie on the same line. Use the table to determine the equation of the line in the form $f(x) = mx + b$.

18.

x	y
−2	5
0	2
2	−1
4	−4
6	−7

19.

x	y
−3	−2
1	3
5	8
9	13
13	18

Use the information displayed in the graph to determine the equation of the line in the form $f(x) = mx + b$.

20.

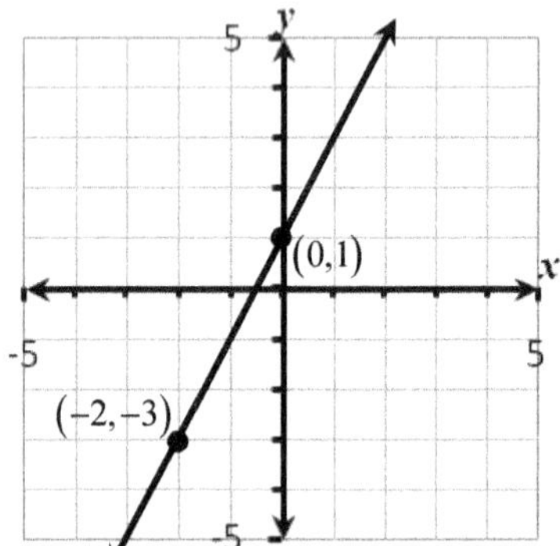

21.

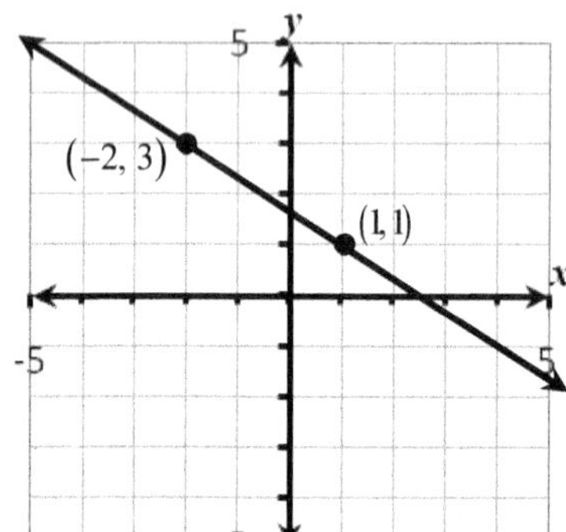

Objective 4. Determine the intercepts of a line.

22. Use the graph to determine the intercepts of the linear function.

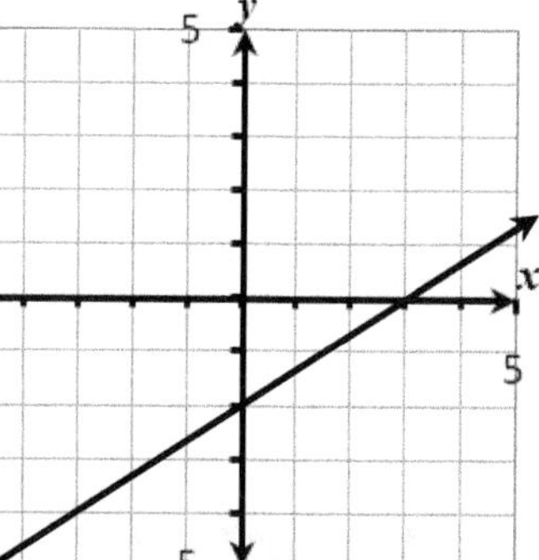

23. Use the table to determine the intercepts of the linear function.

x	y
–3	5
–2	0
–1	–5
0	–10
1	–15
2	–20
3	–25

24. Determine the intercepts of the linear function. Then use the intercepts to sketch a graph of the function.

$$f(x) = -\frac{2}{3}x + 6$$

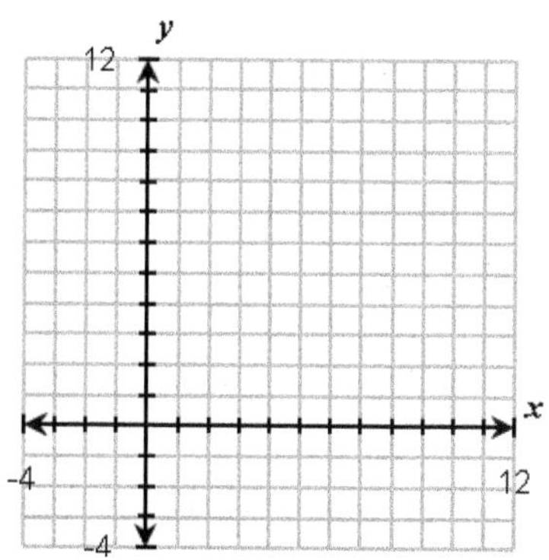

Objective 5: Determine the x-values for which a linear function is positive and the x-values for which a linear function is negative.

Classifying a Function as Positive or Negative			
Verbally	**Graphically**	**Numerically**	**Algebraically**
The function is **positive** at (x, y).	(x, y) is ________ the x-axis.	(x, y) has a positive y-value.	$f(x) > 0$
The function is **negative** at (x, y).	(x, y) is ________ the x-axis.	(x, y) has a negative y-value.	$f(x) < 0$

Determine the x-values for which each linear function is positive and the x-values for which each function is negative.

25.

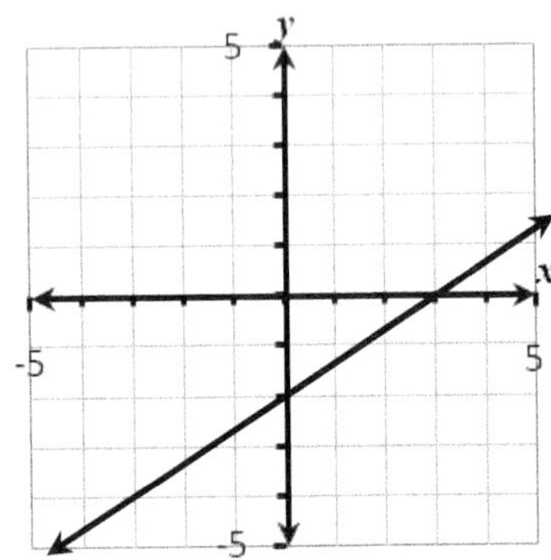

26.

x	y
−3	5
−2	0
−1	−5
0	−10
1	−15
2	−20
3	−25

27. $f(x) = -\frac{1}{3}x + 2$

28. Determine the profit and loss intervals for the profit function graphed below. The x-variable represents the number of units of production and y-variable represents the profit generated by the sale of this production.

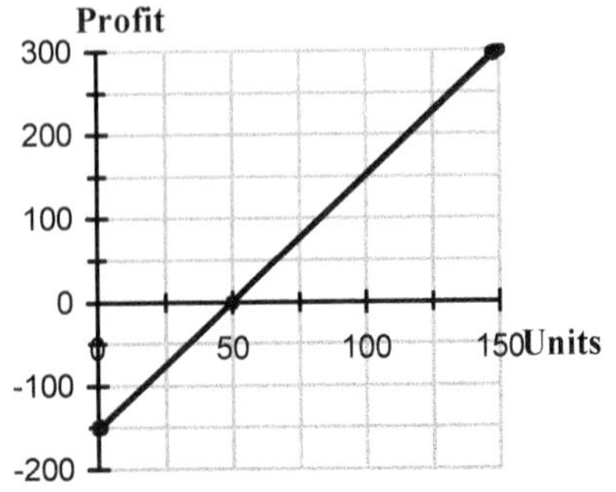

Profit interval:

Loss interval:

29. Write in slope-intercept form the equation of a line passing through $(-3,1)$ and perpendicular to $y=-\frac{3}{4}x-5$.

30. Use the function $f(x)=-4x+8$ to determine the missing input and output values.

(a) $f(4)=$______ **(b)** $f(x)=4;\ x=$______

31. Temperature Beginning at 6:00 am, when the temperature was $38°$ Fahrenheit, the temperature increased by $8°$ per hour over the next 6 hours.

(a) Write a linear function T so that $T(x)$ gives the temperature in degrees Fahrenheit after x hours.

(b) Determine the temperature at noon.

(c) Use this function to determine how many hours until the temperature reaches $74°$.

(d) Use this function to complete the table.

x	$T(x)$
0	
1	
2	
3	
4	
5	
6	

8.3 Lecture Guide: Absolute Value Functions

Objective 1: Sketch the graph of an absolute value function.

Absolute Value Function

Algebraically	**Numerically**	**Graphically**	**Verbally**
$f(x)=\|x\|$	x | $f(x)=\|x\|$ -3 | 3 -2 | 2 -1 | 1 0 | 0 1 | 1 2 | 2 3 | 3	$f(x)=\|x\|$ Vertex 5, -5, x, y	This V-shaped graph opens upward with the vertex of the V-shape at $(0,0)$. The minimum y-value is 0. The domain, the projection of this graph onto the x-axis, is $D=\mathbb{R}$. The range, the projection of this graph onto the y-axis, is $R=[0,\infty)$. The function is decreasing for $x<0$ and increasing for $x>0$.

Complete the table and graph each absolute value function.

1. Equation: $f(x)=-|x|+2$

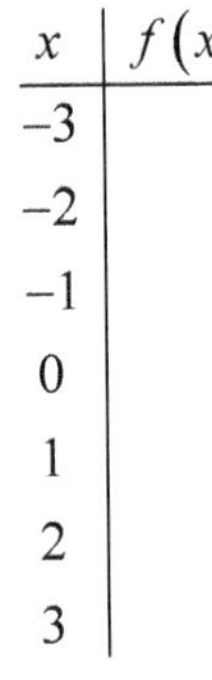

x	$f(x)$
-3	
-2	
-1	
0	
1	
2	
3	

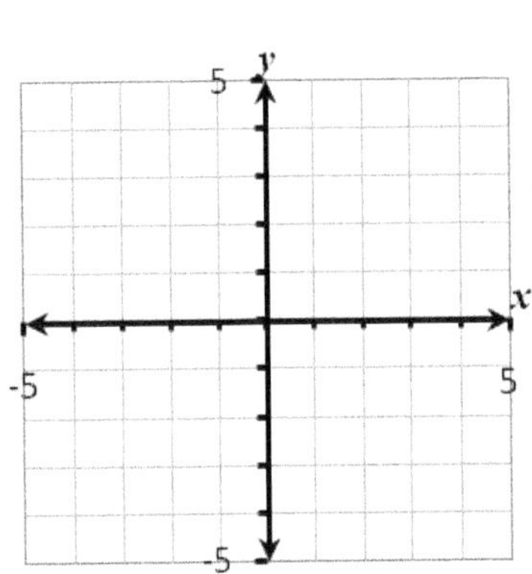

(a) Does the graph open up or down?

(b) Determine the vertex.

(c) Determine the maximum or minimum y-value.

(d) Determine the domain.

(e) Determine the range.

2. Equation: $f(x)=|x-2|$

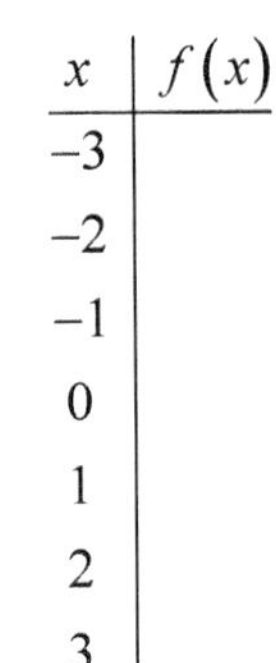

x	$f(x)$
-3	
-2	
-1	
0	
1	
2	
3	

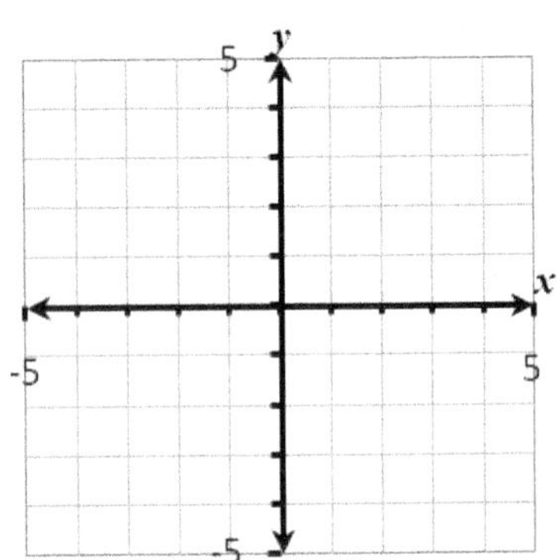

(a) Does the graph open up or down?

(b) Determine the vertex.

(c) Determine the maximum or minimum y-value.

(d) Determine the domain.

(e) Determine the range.

Complete the table, graph the absolute value function, and determine the following.

3. $f(x) = -|x-1| + 2$

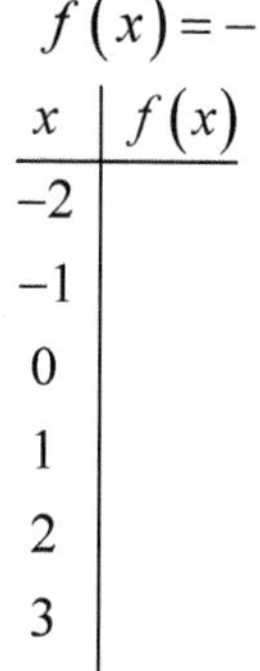

x	$f(x)$
-2	
-1	
0	
1	
2	
3	
4	

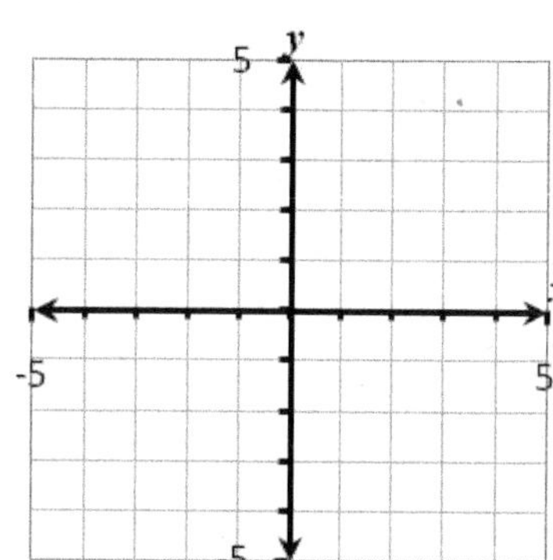

(a) Vertex

(b) The maximum or minimum y-value

(c) Domain

(d) Range

Sketching the Graph of an Absolute Value Function

1. Determine whether the graph opens upward or opens downward.
   - Absolute value functions of the form $f(x) = |ax+b| + c$ will open ____________________.
   - Absolute value functions of the form $f(x) = -|ax+b| + c$ will open ____________________.
2. Determine the location of the vertex.
   - The x-coordinate of the vertex can be found be setting the expression inside the absolute value equal to zero and solving for x.
   - The y-coordinate of the vertex can be found by evaluating the function at this x-coordinate.
3. Complete a table of values using x-values on both sides of the vertex.

Sketch the graph of each absolute value function. Use a calculator only as a check.

4. $f(x) = |2x-3| - 4$

(a) Will this graph open upward or downward?

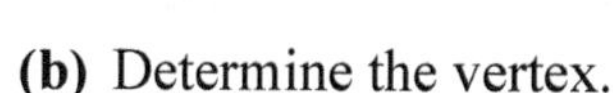

(b) Determine the vertex.

(c) Complete the table and the graph.

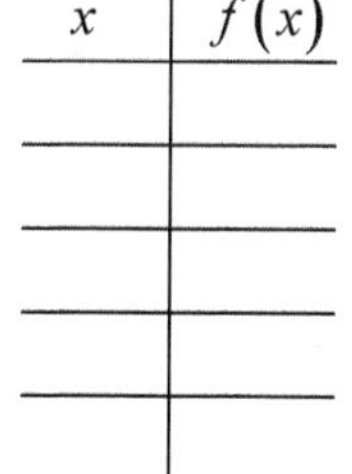

x	$f(x)$

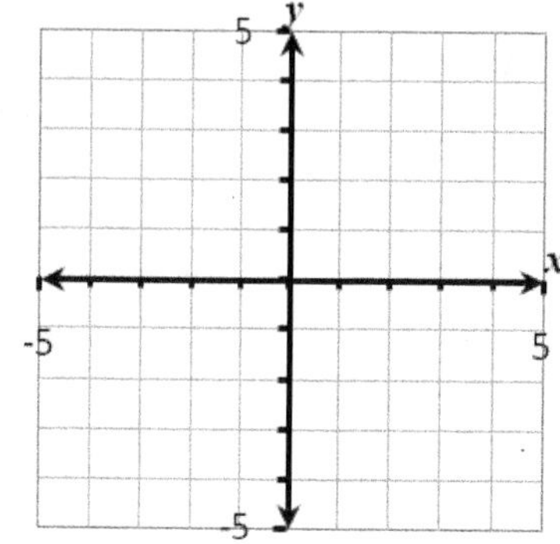

Sketch the graph of each absolute value function. Use a calculator only as a check.

5. $f(x) = -|3x-5|+1$

(a) Will this graph open upward or downward?

(b) Determine the vertex.

(c) Complete the table and the graph.

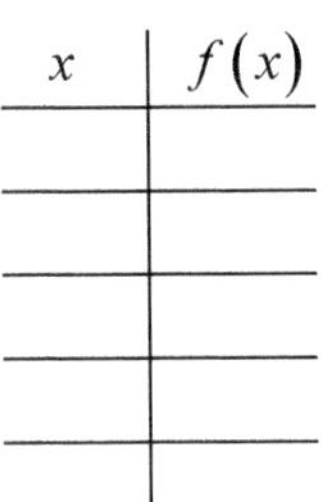

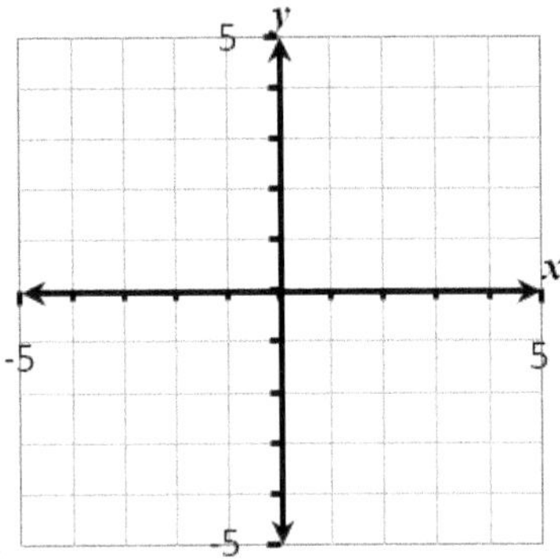

Objective 2. Determine the intercepts of the graph of an absolute value function.

Absolute Value Equations and Inequalities

For any real numbers x and a and positive real number d:

Absolute Value Expression	**Verbally**	**Graphically**	**Equivalent Expression**
$\lvert x-a\rvert = d$	x is either d units left or right of a.	$a-d$ a $a+d$	$x-a=-d$ or $x-a=+d$
$\lvert x-a\rvert < d$	x is less than d units from a.	$a-d$ a $a+d$	$-d < x-a < +d$
$\lvert x-a\rvert > d$	x is more than d units from a.	$a-d$ a $a+d$	$x-a>d$ or $x-a<-d$

Similar statements can also be made about the order relations less than or equal to $(\leq)$ and greater than or equal to $(\geq)$. Expressions with d negative are examined in the group exercises at the end of this section.

Solve each equation and inequality.

6. $|2x-1| = 5$

7. $|x-6| \leq 7$

8. $|x+4| > 10$

9. $|2x+3| \leq 9$

Determine the x- and y-intercepts of the graph of each absolute value function.

10.

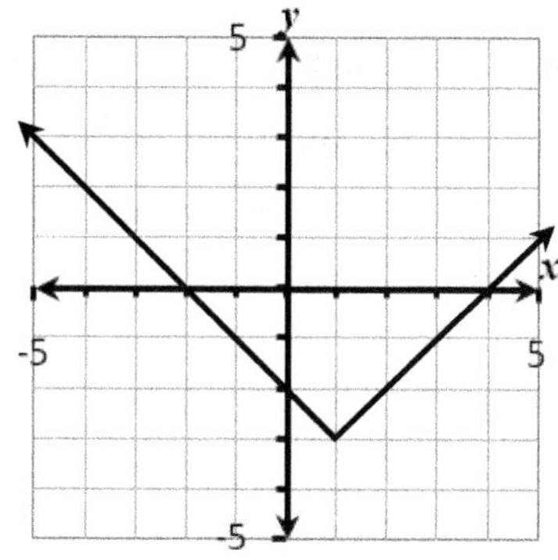

11.

x	y
−3	−4
−2	−2
−1	0
0	2
1	4
2	2
3	0

Algebraically determine the x- and y-intercepts of the graph of each absolute value function.

12. $f(x) = |x - 3| - 4$

13. $f(x) = -|x + 2| + 5$

Objective 3. Determine the x-values for which an absolute value is positive and the x-values for which an absolute value function is negative.

Recall from Section 8.2 that a function $y = f(x)$ is positive when the output value $f(x)$ is positive. On the graph of $y = f(x)$ this occurs at points ____________ the x-axis. The y-values are all positive. A function $y = f(x)$ is negative when the output value $f(x)$ is negative. On the graph of $y = f(x)$ this occurs at points ____________ the x-axis. The y-values are all negative.

Use the graph or table for each absolute value function to determine the interval of x-values for which the function is positive and the interval of x-values for which the function is negative.

14.

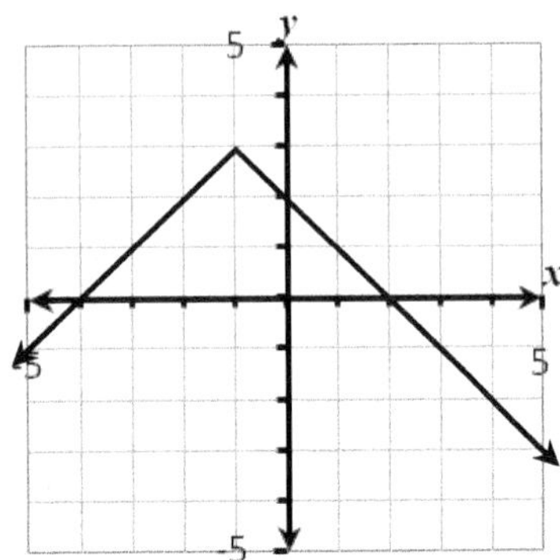

15.

x	y
−3	0
−2	−3
−1	0
0	3
1	6
2	9
3	12

Algebraically determine the interval of x-values for which the function is positive and the interval of x-values for which the function is negative.

16. $f(x) = -|x-2| + 4$

17. $f(x) = |x+1| - 4$

18. Use the function $f(x)=|x+5|-7$ to determine the missing input and output values.

(a) $f(-2)=$ ______

(b) $f(x)=-2;\ x=$ ______ and $x=$ ______

19. Use the function $f(x)=|x+5|-1$ to solve each equation and inequality.

(a) $f(x)=10$

(b) $f(x)<10$

(c) $f(x)\geq 10$

8.4 Lecture Guide: Quadratic Functions

Objective 1: Distinguish between linear and quadratic functions.

A second-degree polynomial function can be written in the form $f(x) = ax^2 + bx + c$ and is called a **quadratic function**.

First Degree Functions --- Linear Functions

Algebraically:
$f(x) = mx + b$

Example:
$f(x) = 2x - 5$

Numerically:
A fixed change in x produces a constant change in y.

Example:

x	y
−3	−11
−2	−9
−1	−7
0	−5
1	−3
2	−1
3	1

Graphically:
A straight line

Example:

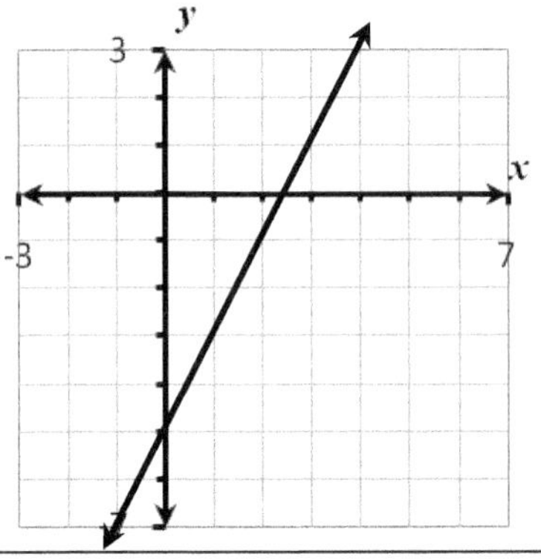

Second Degree Functions --- Quadratic Functions

Algebraically:
$f(x) = ax^2 + bx + c$

Example:
$f(x) = x^2 - 4x + 4$

Numerically:
The y-values from a symmetric pattern about the vertex.

Example:

x	y
−1	9
0	4
1	1
2	0
3	1
4	4
5	9

Graphically:
A parabola

Example:

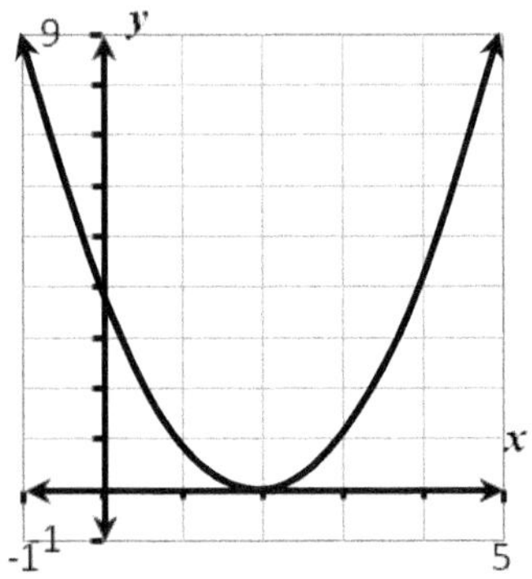

1. For a linear function in the form $f(x) = mx + b$, the graph will slope upward to the right if __________________ and the graph will slope downward to the right if __________________.

2. For quadratic functions in the form $f(x) = ax^2 + bx + c$, the graph will open up if __________________ and the graph will open down if __________________.

Identify the graph of each function as a line or a parabola. If the graph is a line, determine whether the slope is negative or positive. If the graph is a parabola, determine whether the parabola is concave up (the graph opens up) or concave down (the graph opens down).

3. $y = -3x + 8$

4. $y = 2x^2 + 9x + 4$

5. $y = -3x^2 + 7x + 6$

6. $y = 9x - 5$

Objective 2: Determine the vertex of a parabola.

For a parabola defined by $f(x) = ax^2 + bx + c$, the x-intercepts (if they exist) can be determined by setting $y = 0$ and then using the quadratic formula. The vertex will be located at the x-value midway between the two x-intercepts. See the figure below.

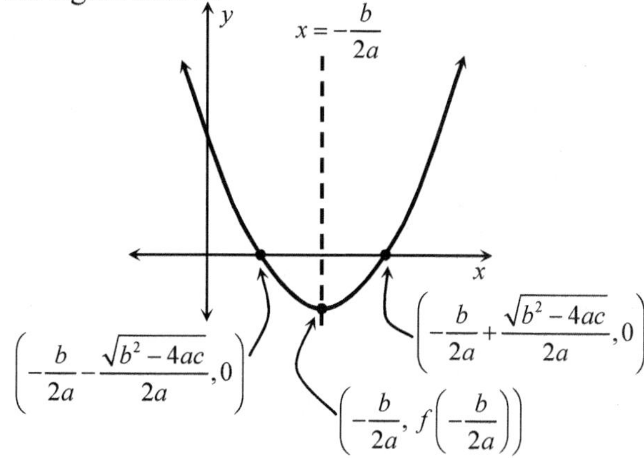

7. The x-intercepts of a parabola are $(-5, 0)$ and $(8, 0)$. Determine the x-coordinate of the vertex.

8. The x-intercepts of a parabola are $(7, 0)$ and $(15, 0)$. Determine the x-coordinate of the vertex.

Vertex of the Parabola Defined by $f(x) = ax^2 + bx + c$

Algebraically

$\left(-\frac{b}{2a}, f\left(-\frac{b}{2a}\right)\right)$

Example:

$f(x) = -x^2 + 7x + 8$

$a = -1$ and $b = 7$

$x = -\frac{b}{2a} = -\frac{7}{2(-1)} = \frac{7}{2} = 3.5$

$f(3.5) = 20.25$

The vertex is $(3.5, 20.25)$.

Numerically

The y-values form a symmetric pattern about the vertex. If the table contains the vertex, the y-coordinate of the vertex will be either the largest or the smallest y-value in the table.

Example:

x	y
-2	-10
-1	0
2	18
3.5	20.25
5	18
8	0
9	-10

Graphically

The vertex is either the highest or the lowest point on the parabola.

Example:

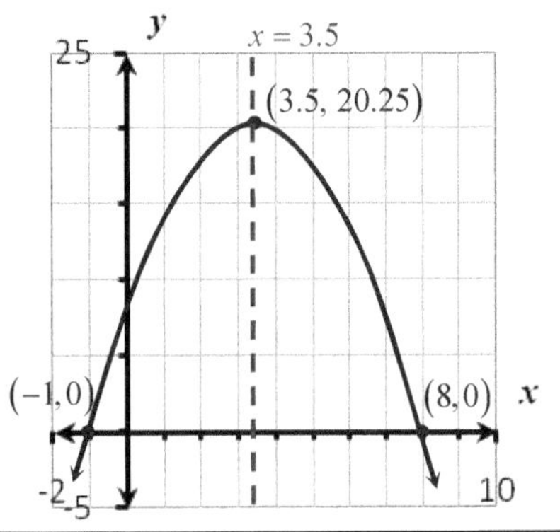

Finding the Vertex of a Parabola Defined by $f(x) = ax^2 + bx + c$

Step 1. Determine the x-coordinate using $x = -\frac{b}{2a}$.

Step 2. Then evaluate $f\left(-\frac{b}{2a}\right)$ to determine the y-coordinate.

9. Determine the vertex of the parabola defined by $f(x) = 3x^2 + 12x - 1$.

10. Determine the vertex of the parabola defined by $f(x) = -2x^2 + 6x + 5$.

Use the given equation to calculate the x and y-intercepts and the vertex of each parabola.

11. $y = 2x^2 - 5x - 3$

(a) y-intercept

(b) x-intercepts

(c) Vertex

12. $y = (2x - 7)(x + 1)$

(a) y-intercept

(b) x-intercepts

(c) Vertex

Objective 3: Sketch the graph of a quadratic function and determine key features of the resulting parabola.

Complete the table, plot the points on the graph, and connect these points with a smooth parabolic curve. Then complete the missing information.

13. $f(x) = -x^2 + 2x + 8$

Open upward or downward?

Vertex:

***x*-intercepts:**

***y*-intercept:**

Domain:

Range:

Table

x	y
−2	
−1	
0	
1	
2	
3	
4	

Graph

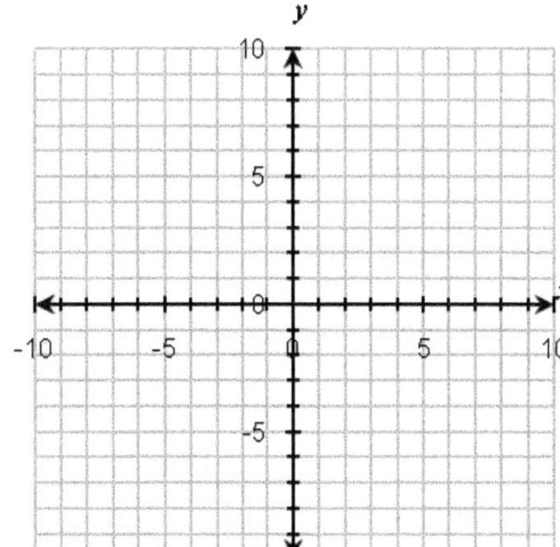

14. $f(x) = x^2 + 2x - 3$

Open upward or downward?

Vertex:

***x*-intercepts:**

***y*-intercept:**

Domain:

Range:

Table

x	y
−4	
−3	
−2	
−1	
0	
1	
2	

Graph

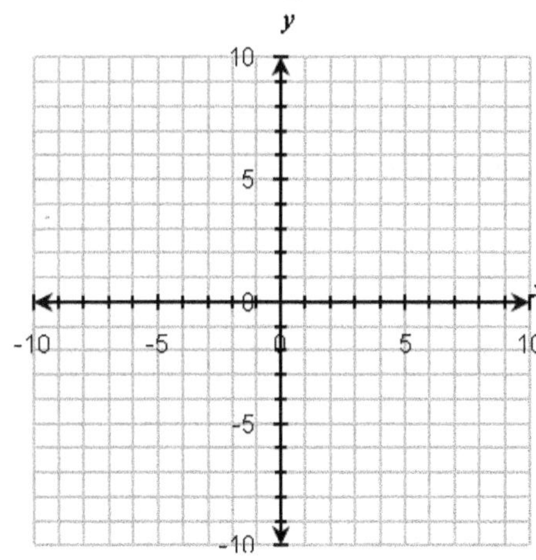

Sketching the Graph of a Quadratic Function

1. Determine whether the parabola opens upward or downward.
2. Determine the coordinates of the vertex.
3. Determine the intercepts.
4. Complete a table using points on both sides of the vertex.
5. Connect all points with a smooth parabolic shape.

15. Sketch the graph of $f(x) = x^2 + 2x - 8$.

(a) Will the parabola open upward or open downward?

(b) Determine the coordinates of the vertex.

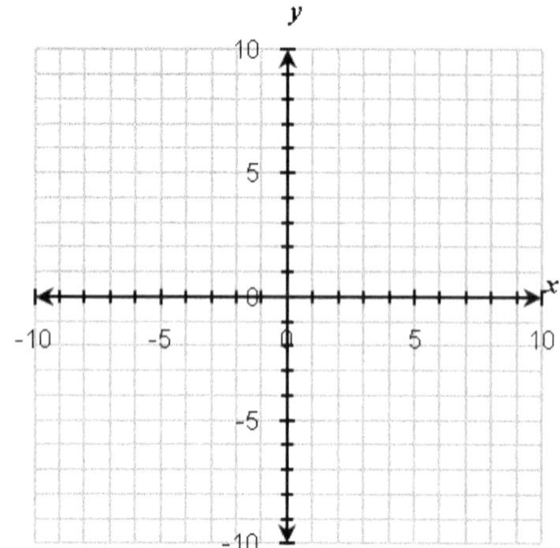

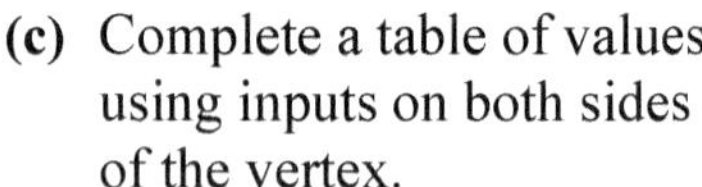

(c) Complete a table of values using inputs on both sides of the vertex.

(e) Use this information to sketch the graph of this function.

(d) Determine the intercepts of the graph of this function.

16. Sketch the graph of $f(x) = -2x^2 - 11x + 6$.

(a) Will the parabola open upward or open downward?

(b) Determine the coordinates of the vertex.

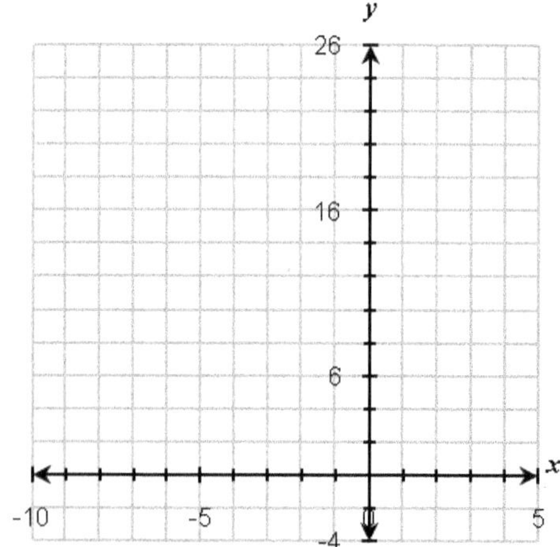

(c) Complete a table of values using inputs on both sides of the vertex.

(e) Use this information to sketch the graph of this function.

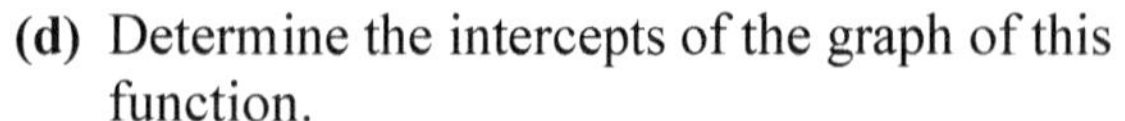

(d) Determine the intercepts of the graph of this function.

Objective 4: Solve problems involving a maximum or minimum value.

17. $P(x)$ gives the profit in dollars when x units are produced and sold. Use the graph of the profit function to determine the following:

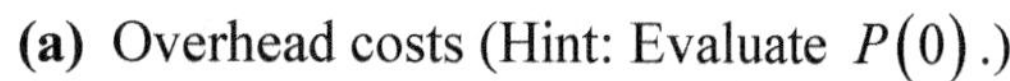

(a) Overhead costs (Hint: Evaluate $P(0)$.)

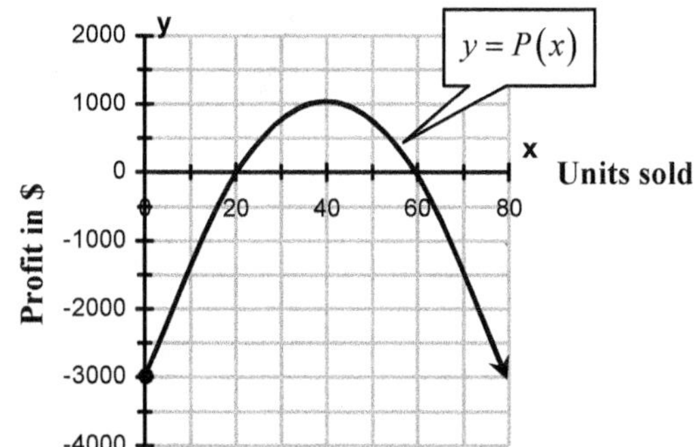

(b) Break even values (Hint: When does $P(x) = 0$?)

(c) Maximum profit that can be made and the number of units to sell to create this profit.

18. A rancher has 240 yards of fencing available to enclose 3 sides of a rectangular corral. A river forms one side of the corral

(a) If x yards are used for the two parallel sides, how much fencing remains for the side parallel to the river? Give this length in terms of x.

$L =$ ________________

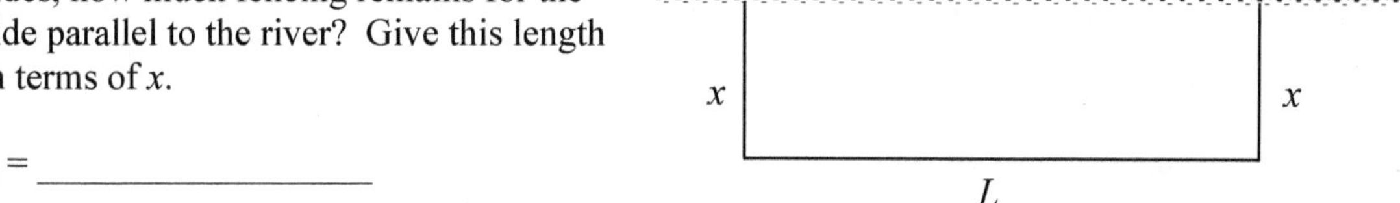

(b) Express the total area of the fenced corral as a function of x. Hint: Area = (Length)(Width)

$A(x) =$ ________________

(c) What is the maximum area that can be enclosed with this fencing?

Maximum area = ________________

19. The equation $y = -16x^2 + 80x + 3$ gives the height y of a baseball in feet x seconds after it was hit.

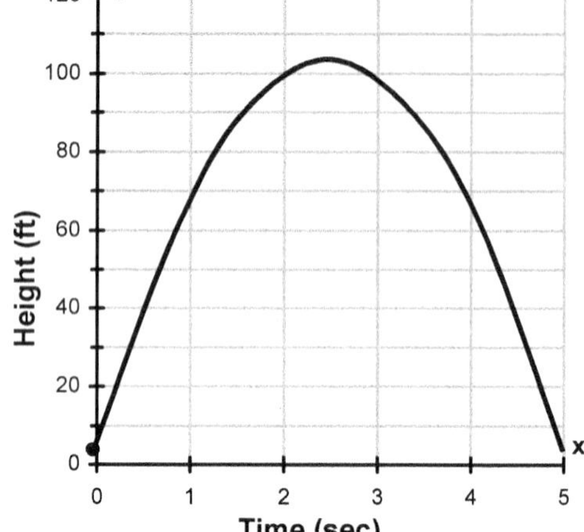

(a) Use the equation to determine how many seconds into the flight the maximum height is reached.

(b) Determine the maximum height the ball reached.

(c) Do your results agree with what you can observe from the graph?

Use your graphing calculator to determine the minimum/maximum value of $f(x)$ and the x-value at which this minimum/maximum occurs. Use a window of $[-10,10,1]$ by $[-50,50,5]$ for each graph. See Technology Perspective 8.4.1.

20. $y = x^2 + 4x - 21$

Sketch of calculator graph:

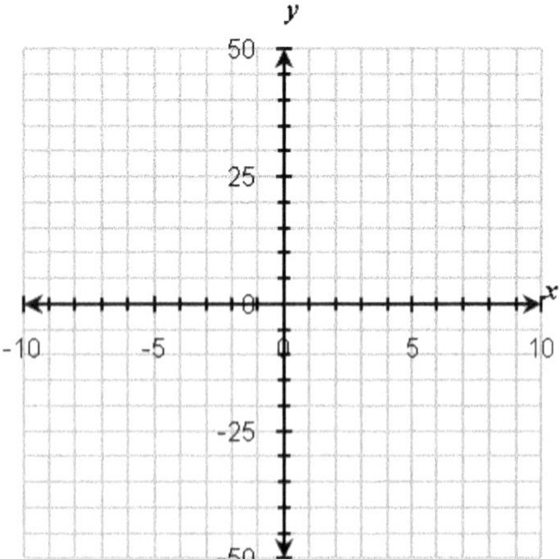

Max/min value:

***x*-value where max/min occurs:**

21. $y = -x^2 + 5x + 36$

Sketch of calculator graph:

y

50

25

0

-25

-50

x

-10 -5 5 10

Max/min value:

***x*-value where max/min occurs:**

22. Use the function $f(x)=x^2-4x-15$ to solve each equation and inequality.

(a) $f(x)=6$

(b) $f(x)<6$

(c) $f(x)\geq 6$

23. Use the given graph to determine the missing input and output values.

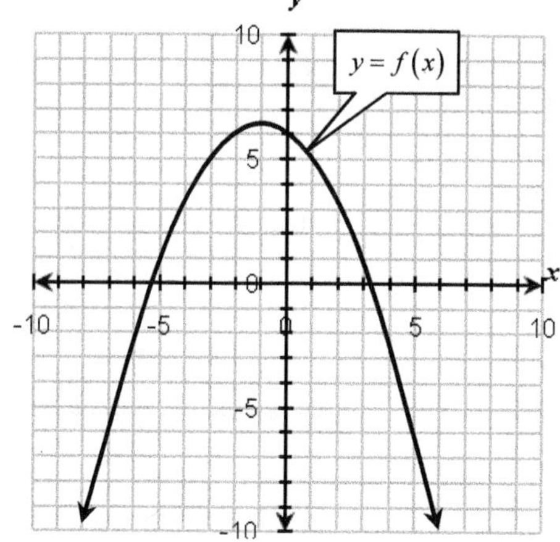

(a) $f(1)=$ ______

(b) $f(x)=1$; $x=$ ______ and $x=$ ______

24. Use the function $f(x)=-x^2-5x+16$ to determine the missing input and output values.

(a) $f(2)=$ ______

(b) $f(x)=2$; $x=$ ______ and $x=$ ______

8.5 Lecture Guide: Analyzing Graphs

Objective 1: Identify a function as an increasing or decreasing function.

Increasing and Decreasing Functions

	Graphical Example	Verbally
Increasing Function:		A function is increasing over its entire domain if the graph rises as it moves from left to right. For all x-values, as the x-values increase, the y-values also increase.
Decreasing Function:		A function is decreasing over its entire domain if the graph drops as it moves from left to right. For all x-values, as the x-values increase, the y-values decrease.

Use the graph of each function to identify the function as an increasing function, a decreasing function, or a function that is neither increasing nor decreasing.

1.

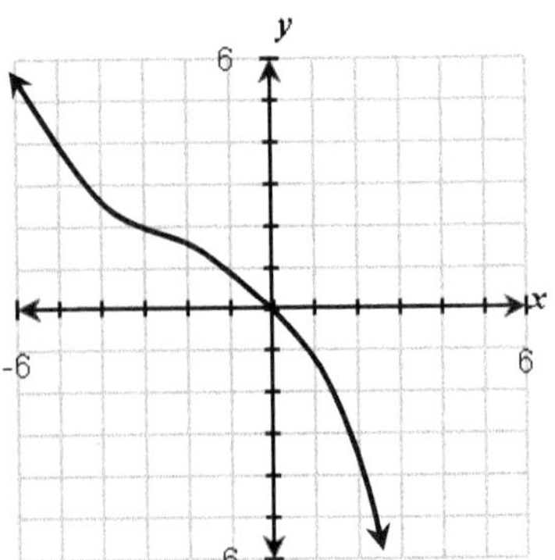

2.

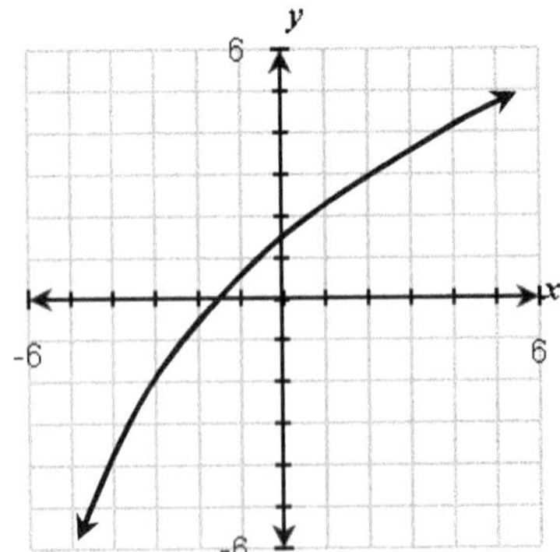

3.

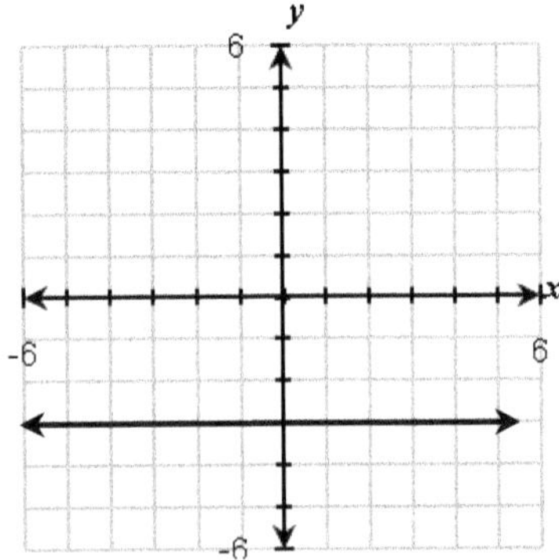

Classification of Slopes

Positive Slope: An Increasing Linear Function	**Negative Slope: A Decreasing Linear Function**	**Zero Slope: A Constant Linear Function**
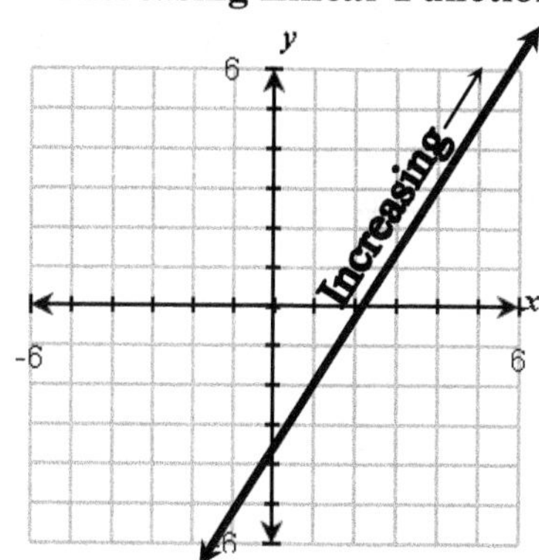	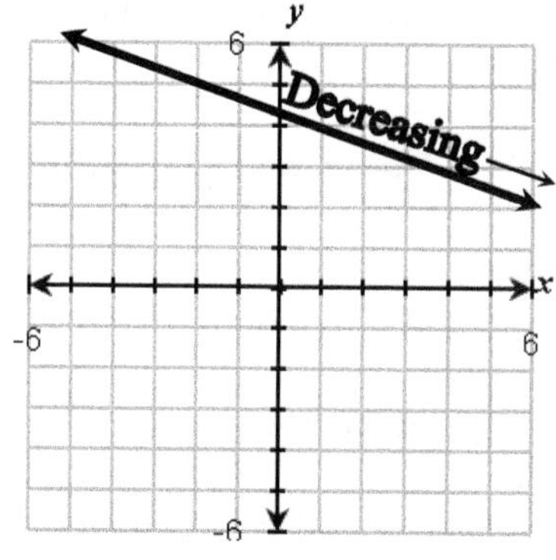	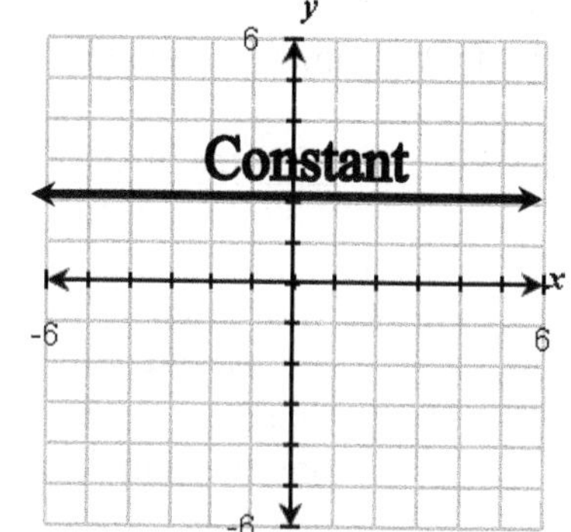
For an increasing function, a positive value of Δx produces a ___________ value of Δy.	For a decreasing function, a positive value of Δx produces a ___________ value of Δy.	For a constant function, any change in x produces ______ change in y.

4. For a linear function in the form $f(x) = mx + b$:

(a) The function is increasing if ________________.

(b) The function is decreasing if ________________.

(c) The function is constant (neither increasing nor decreasing) if ________________.

Use the equation defining each function to identify the function as an increasing function, a decreasing function, or a constant function. Use a graphing calculator only as a check.

5. $f(x) = 5x - 7$ **6.** $f(x) = -10$ **7.** $f(x) = -3x + 12$

8. The graph gives the altitude y in thousands of feet in terms of the time x in minutes since the start of a flight in a small plane.

(a) During what time interval was the altitude increasing?

(b) During what time interval was the altitude constant?

(c) During what time interval was the altitude decreasing?

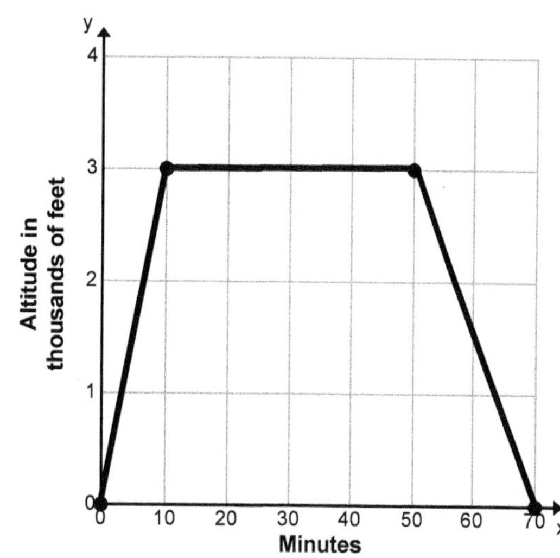

Objective 2: Analyze the graphs of linear, absolute value, and quadratic functions.

The following box contains a check list of key information that we are currently able to obtain about the graphs of linear, absolute value, and quadratic functions. We will use this list as a guide for analyzing each of these functions and their graphs.

Analyzing the Graphs of Linear, Absolute Value, and Quadratic Functions

1. Identify the type of function.
   - **a.** For linear functions, determine the slope.
   - **b.** For absolute value and quadratic functions, determine the vertex and whether the graph opens upward or downward.
2. Determine the y-intercept and the x-intercept(s).
3. Determine the domain and range.
4. Determine where the function is positive and where the function is negative.
5. Determine where the function is increasing and where the function is decreasing.

9. Use the graph of the linear function to determine the missing information.

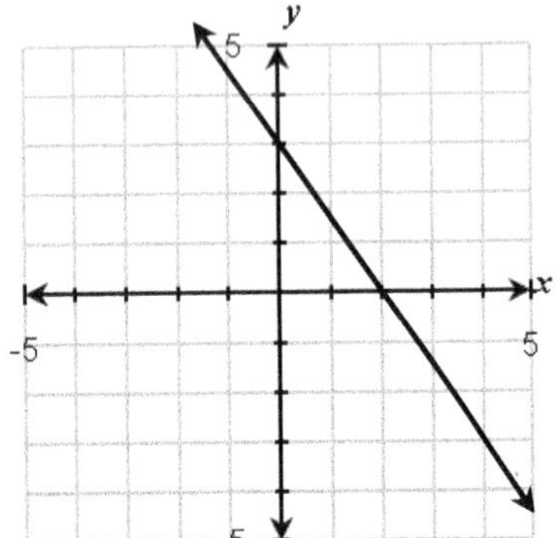

(a) Determine the slope of this line.

(b) Determine the x-intercept.

(c) Determine the y-intercept.

(d) Determine the domain of this function.

(e) Determine the range of this function.

(f) Determine the x-values for which the function is positive.

(g) Determine the x-values for which the function is negative.

(h) Determine the x-values for which the function is decreasing.

(i) Determine the x-values for which the function is increasing.

10. Use the linear function $f(x) = -25x + 300$ to determine the missing information.

(a) Determine the slope of this line.

(b) Determine the x-intercept.

(c) Determine the y-intercept.

(d) Determine the domain of this function.

(e) Determine the range of this function.

(f) Determine the x-values for which the function is positive.

(g) Determine the x-values for which the function is negative.

(h) Determine the x-values for which the function is decreasing.

(i) Determine the x-values for which the function is increasing.

11. Use the graph of the absolute value function to determine the missing information.

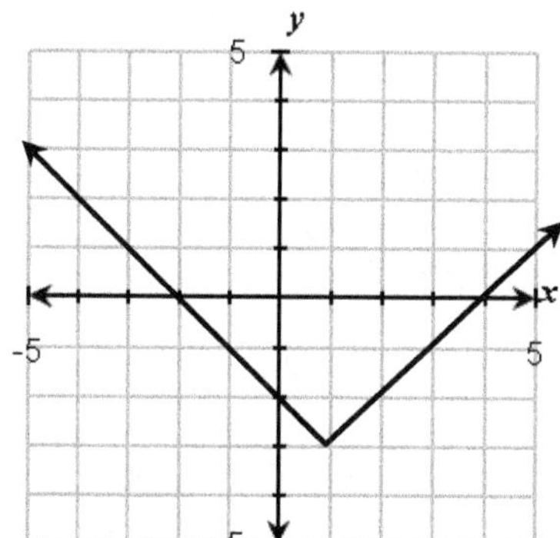

(a) Determine the vertex of this graph.

(b) Does the graph open upward or downward?

(c) Determine the x-intercept(s).

(d) Determine the y-intercept.

(e) Determine the domain of this function.

(f) Determine the range of this function.

(g) Determine the x-values for which the function is positive.

(h) Determine the x-values for which the function is negative.

(i) Determine the x-values for which the function is decreasing.

(j) Determine the x-values for which the function is increasing.

12. Use the absolute value function $f(x) = -|x-20|+12$ to determine the missing information.

(a) Determine the vertex of this graph.

(b) Does the graph open upward or downward?

(c) Determine the x-intercept(s).

(d) Determine the y-intercept.

(e) Determine the domain of this function.

(f) Determine the range of this function.

(g) Determine the x-values for which the function is positive.

(h) Determine the x-values for which the function is negative.

(i) Determine the x-values for which the function is decreasing.

(j) Determine the x-values for which the function is increasing.

13. Use the graph of the quadratic function to determine the missing information.

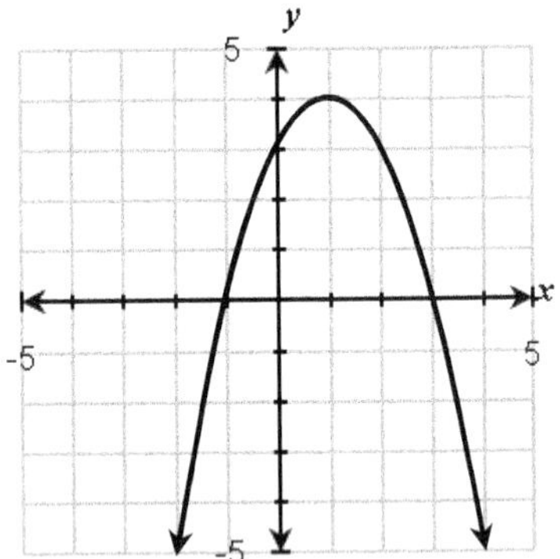

(a) Determine the vertex of this graph.

(b) Does the graph open upward or downward?

(c) Determine the x-intercept(s).

(d) Determine the y-intercept.

(e) Determine the domain of this function.

(f) Determine the range of this function.

(g) Determine the x-values for which the function is positive.

(h) Determine the x-values for which the function is negative.

(i) Determine the x-values for which the function is decreasing.

(j) Determine the x-values for which the function is increasing.

14. Use the quadratic function $f(x)=x^2-7x-30$ to determine the missing information.

(a) Determine the vertex of this graph.

(b) Does the graph open upward or downward?

(c) Determine the x-intercept(s).

(d) Determine the y-intercept.

(e) Determine the domain of this function.

(f) Determine the range of this function.

(g) Determine the x-values for which the function is positive.

(h) Determine the x-values for which the function is negative.

(i) Determine the x-values for which the function is decreasing.

(j) Determine the x-values for which the function is increasing.

15. The height in feet of a ball thrown off the roof of a building is given by $h(t) = -16t^2 + 60t + 75$, where t represents the time in seconds since the ball was thrown. This function is also represented in the graph and table below. Answer each question below algebraically. Use the graph and the table to verify your results.

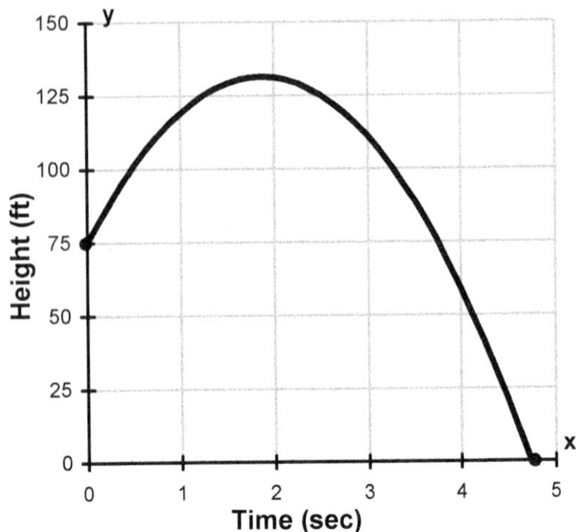

Time (sec)	Height (ft)
0.0	75
0.5	101
1.0	119
1.5	129
2.0	131
2.5	125
3.0	111
3.5	89
4.0	59
4.5	21

(a) Determine the y-intercept and interpret its meaning.

(b) Determine the x-intercepts and determine their meaning.

(c) Determine the vertex and interpret the meaning of each coordinate.

(d) Determine the t-values for which the function is increasing.

(e) Interpret the meaning of the interval in part **(d)**.

8.6 Lecture Guide:
Curve Fitting – Selecting a Line or Parabola that Best Fits a Set of Data

Objective 1: Create a scatter diagram and select an appropriate model for the data.

Draw a scatter diagram for each set of data points and then select the shape that best describes the pattern formed by these points.

A. A line with positive slope
B. A parabola opening upward
C. A line with negative slope
D. A parabola opening downward

1.

x	−3	−2	−1	0	1
y	8	5	4	5	8

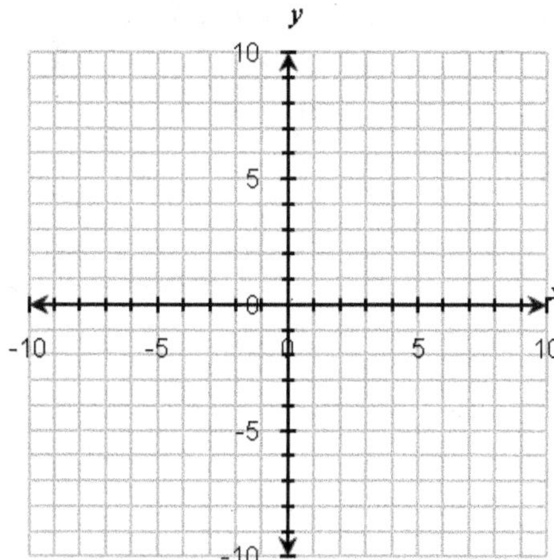

2.

x	−3	−2	−1	0	1
y	8	5	2	−1	−4

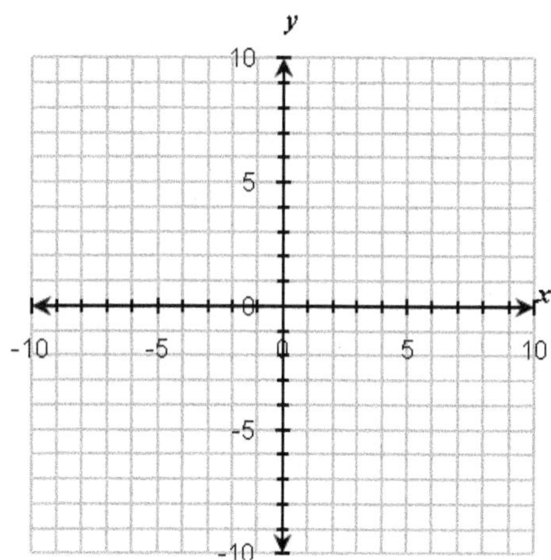

Objective 2: Select a linear function that best fits a set of data.

To use a graphing calculator to draw a scatter diagram for a set of data points

x	−3	−2	−1	0	1	2	3
y	−9	−7	−5	−3	−1	1	3

and determine the line of best fit, first enter the data points in the lists **L1** and **L2**. Follow this procedure to enter the data points in the lists.

TI-84 Plus Keystrokes

1. To **Edit** lists **L1** and **L2**,

   press  .

2. Enter the x-values under **L1**.

3. Press [⟩] to access **L2**. Enter the y-values under **L2**.

TI-84 Plus Screens

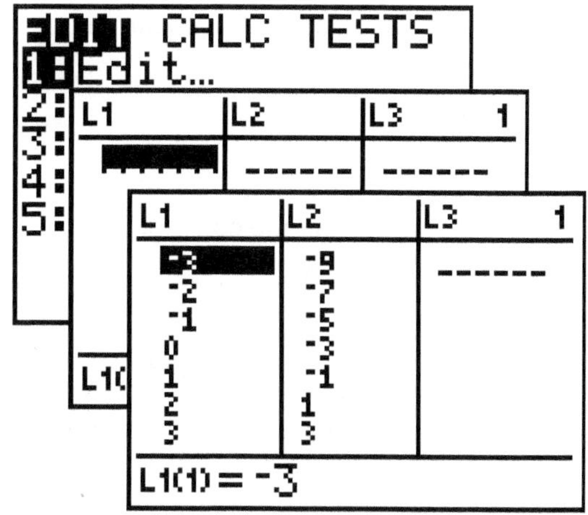

To view the scatter diagram, first make sure Plot1 is turned on. Note that Plot1 is highlighted at the top of the **Y=** screen when it is turned on. Follow this procedure to view the scatter diagram.

TI-84 Plus Keystrokes | **TI-84 Plus Screens**

1. Turn on **Plot1** by pressing 2nd Y= . Set up the menu as shown here.

2. Press Y= and check to see that the **Y=** screen is blank.

3. Use **Zoom Stat** to automatically select a graphing window for these points. This is option 9 in the ZOOM menu.

$[-3.6, 3.6, 1]$ by $[-11.04, 5.04, 1]$

Note: This calculator display corresponds to the linear pattern shown in Example 1-a.

Follow this procedure to view the scatter diagram to calculate the line of best fit for the data and then to graph this line.

TI-84 Plus Keystrokes | **TI-84 Plus Screens**

1. Select the **LinReg** option --- this is option 4 in the **STAT-CALC** menu --- by pressing

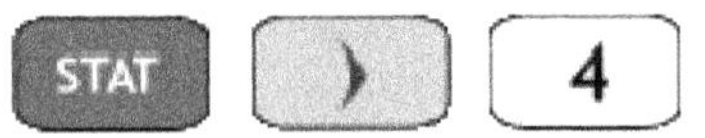

2. Press ENTER to calculate the equation of the line of best fit.

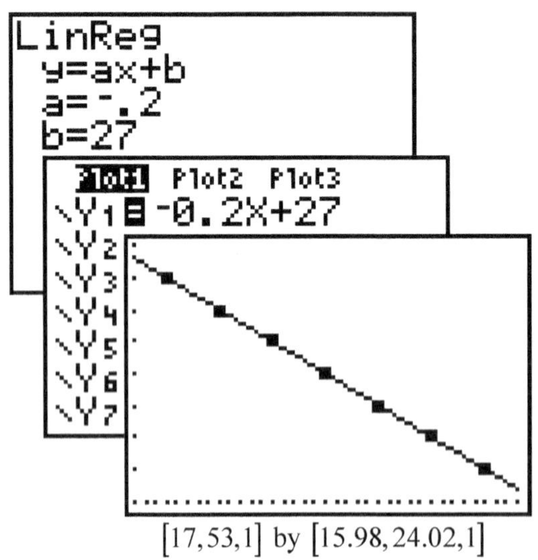

$[17, 53, 1]$ by $[15.98, 24.02, 1]$

3. Press Y= and enter the equation of the line of best fit as shown.

4. Use **Zoom Stat** --- option 9 in the ZOOM menu --- to display the data points along with the graph of the line of best fit.

Use a graphing calculator to draw a scatter diagram for the data points and determine the line of best fit. Express your answer in the form $y = mx + b$ with m and b rounded to the nearest thousandth if necessary.

3.

x	y
1	−2.1
2	0.4
4	5.8
5	8.7
6	11.4

(a) Line of best fit:

(b) Use this linear equation to estimate y when $x = 3$

4.

x	y
3	20
6	37
9	48
12	59
18	88

(a) Line of best fit:

(b) Use this linear equation to estimate y when $x = 11$

5.

x	y
−4	1.5
−1	2.2
2	2.9
5	4.1
8	4.5

(a) Line of best fit:

(b) Use this linear equation to estimate y when $x = 5$

6.

x	y
5	290
10	110
15	−90
20	−220
25	−340

(a) Line of best fit:

(b) Use this linear equation to estimate y when $x = 10$

7. At a wood shop, there are oak boards of various sizes. The following table shows the prices and lengths of some of these boards.

x length (ft)	y cost ($)
2	3.25
5	8.25
8	13.00
10	16.25

(a) Determine the equation of the line of best fit for this data. Round to the nearest thousandth.

(b) Give the slope of the line of best fit and interpret its meaning.

(c) Use this equation to estimate the cost of a 6-foot board.

Objective 3: Select a quadratic function that best fits a set of data.

8. There is only one keystroke that is different for finding the parabola of best fit compared to finding the line of best fit. The keystrokes required to calculate the parabola of best fit are:

______ ______ ______ ______

Use a graphing calculator to draw a scatter diagram of each set of data points. Then determine the quadratic function that best fits these points. Write the equation in the form $y = ax^2 + bx + c$ and round the coefficients to the nearest thousandth if necessary.

9.

x	y
−8	28
−3	10
1	5
6	15
9	27

(a) Equation of best fit:

(b) Use this equation to approximate y when $x = 3$

10.

x	y
10	−9.3
20	−6.1
30	−4.8
40	−5.9
50	−8.8

(a) Equation of best fit:

(b) Use this equation to approximate y when $x = 34$

11. For a science experiment, the height in meters of a projectile has been recorded in the table below at 1-second intervals.

(a) Determine the equation of the parabola of best fit for this data.

x time (sec)	y height (m)
1	25
2	30
3	26
4	12

(b) Use this equation to estimate the height of the projectile after 2.5 seconds.

Lecture Guides for

Chapter 9

Rational Functions

9.1 Lecture Guide: Graphs of Rational Functions and Reducing Rational Expressions

Objective 1: Identify rational functions.

Rational Function

Algebraically	**Verbally**	**Algebraic Example**
$f(x)=\dfrac{P(x)}{Q(x)}$ is a rational function if $P(x)$ and $Q(x)$ are ______________ and $Q(x)\neq 0$.	A rational function is defined as the ____________ of two polynomials.	$f(x)=\dfrac{1}{x}$

Determine whether each function defines a rational function.

1. $f(x)=\dfrac{x+1}{x-2}$

2. $f(x)=-2x+5$

3. $f(x)=\lvert x+3\rvert-1$

4. $f(x)=2x^2-3x-4$

5. $f(x)=\sqrt{x-3}+2$

6. $f(x)=\dfrac{x+2}{5}$

Objective 2: Determine the domain and identify the vertical asymptotes of a rational function.

Domain of a Rational Function

The domain of a rational function must exclude values that would cause ______________ by zero.

Algebraic Example

The domain of $f(x)=\dfrac{1}{x}$ is $\mathbb{R}\sim\{0\}$

Numerical Example

x	$y=\dfrac{1}{x}$
-3	-0.333
-2	-0.500
-1	-1.000
0	undefined
1	1.000
2	0.500
3	0.333

Graphical Example

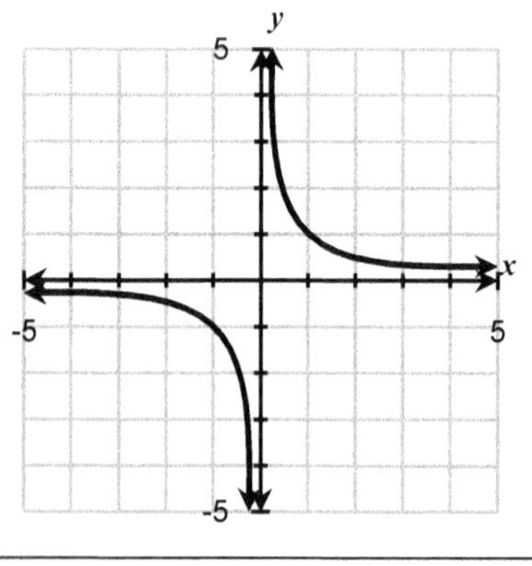

y
5
x
-5
5
-5

Verbal Example

Only zero is excluded from the domain to prevent division by zero. Note that 1 divided by 0 is undefined. Also note that there is a break in the graph at $x=0$.

Characteristics of the graph of $f(x)=\dfrac{1}{x}$

- The domain of this function does not include __________.
- The function is discontinuous, there is a _________ in the graph at $x=0$. The graph consists of two unconnected branches.
- The graph approaches the ______________ line $x=0$ asymptotically.
- The graph approaches the ______________ line $y=0$ asymptotically.

7. If $f(x)=\dfrac{x+4}{x-2}$, evaluate each expression.

(a) $f(0)$

(b) $f(1)$

(c) $f(2)$

8. Given the rational function $f(x)=\dfrac{x+4}{x-2}$:

(a) What value is excluded from the domain of this function?

(b) What is the domain of this function?

(c) What is the equation of the vertical asymptote of this function?

Determine the domain of each rational function.

9. $f(x)=\dfrac{x-1}{x-5}$

10. $f(x)=\dfrac{x+2}{2x-3}$

11. $f(x)=\dfrac{x+4}{x^2-4x+4}$

12. $f(x)=\dfrac{x+5}{x^2-x-6}$

13. Use the given table to determine the domain and vertical asymptotes of $f(x)=\frac{x+7}{x^2+5x+6}$.

x	$f(x)$
−4	1.500
−3	undefined
−2	undefined
−1	3.000
0	1.167
1	0.667
2	0.450

(a) Domain:

(b) Vertical Asymptotes:

14. Use the given graph to determine the domain and vertical asymptotes of $f(x)=\frac{x^2+2}{x^2+4x-5}$.

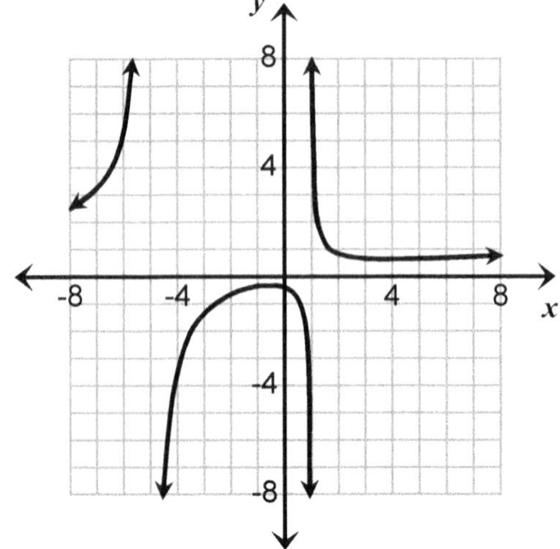

(a) Domain:

(b) Vertical Asymptotes:

Objective 3: Reduce a rational expression to lowest terms.

Reducing a Rational Expression to Lowest Terms

Algebraically	**Verbally**	**Algebraic Example**
If A, B and C are polynomials and $B \neq 0$ and $C \neq 0$ then $\frac{AC}{BC}=\frac{A}{B}$.	**1.** _______________ both the numerator and the denominator of the rational expression. **2.** Divide the numerator and the denominator by any common nonzero _____________.	$\frac{3x-6}{x^2-2x}=\frac{3\overset{1}{\cancel{(x-2)}}}{x\underset{1}{\cancel{(x-2)}}}$ $=\frac{3}{x}$ for $x \neq 2$

Reduce each rational expression.

15. $\frac{3x^4y^6}{12x^7y^4}$

16. $\frac{x^2-49}{3x-21}$

Reduce each rational expression.

17. $\dfrac{3x-5y}{15y-9x}$

18. $\dfrac{7a-7b}{a^2-b^2}$

19. $\dfrac{x^2-4x}{x^2-x-12}$

20. $\dfrac{4x^2-9y^2}{4x^2+12xy+9y^2}$

21. $\dfrac{11-x}{x^2-121}$

22. $\dfrac{2x-3y+7}{-2x+3y-7}$

23. The average cost per jersey for a screen printer to produce x sports jerseys is displayed in this graph.

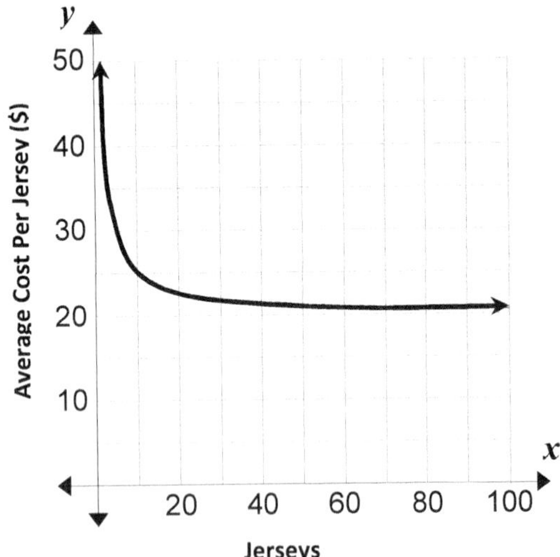

(a) Describe what happens to the average cost per jersey as the number of jerseys produced becomes very small.

(b) Describe what happens to the average cost per jersey as the number of jerseys produced becomes very large.

Fill in the missing numerator or denominator.

24. $\dfrac{x+3}{x-5} = \dfrac{x^2-4x-21}{?}$

25. $\dfrac{x-2}{x+3} = \dfrac{?}{x^2-2x-15}$

9.2 Lecture Guide: Multiplying and Dividing Rational Expressions

Objective 1: Multiply and divide rational expressions.

Multiplying Rational Expressions

Algebraically	**Verbally**	**Algebraic Example**
If A, B, C and D are real polynomials and $B \neq 0$ and $D \neq 0$, then $\frac{A}{B} \cdot \frac{C}{D} = \frac{AC}{BD}$.	**1.** _______________ the numerators and the denominators. Then write the product as a single fraction, indicating the product of the numerators and the product of the denominators. **2.** _______________ this fraction by dividing the numerator and the denominator by any common nonzero factors.	$\frac{x^2 - 5x}{4xy - 12xy} \cdot \frac{4y}{2x - 10}$

Dividing Rational Expressions

Algebraically	**Verbally**	**Algebraic Example**
If A,B, C, and D are real polynomials and $B \neq 0$, $C \neq 0$, and $D \neq 0$ then $\frac{A}{B} \div \frac{C}{D} = \frac{A}{B} \cdot \frac{D}{C}$ $= \frac{AD}{BC}$	**1.** Rewrite the division problem as the product of the dividend and the _______________ of the divisor. **2.** Perform the multiplication using the rule for multiplying rational expressions.	$\frac{20a}{3a - 27} \div \frac{5a}{4a - 36}$

Perform the indicated operations and reduce all results to lowest terms. Assume the variables are restricted to values that prevent division by zero.

1. $\frac{22}{15} \cdot \frac{25}{33}$

2. $\frac{27}{16} \div \frac{9}{32}$

3. $\frac{30x^2}{21y} \cdot \frac{14y^3}{25x^4}$

4. $\frac{26a}{15b} \div \frac{13a^2}{45b}$

Perform the indicated operations and reduce all results to lowest terms. Assume the variables are restricted to values that prevent division by zero.

5. $\frac{x+3}{x^2-4} \cdot \frac{x^2-4x+4}{x^2+x-6}$

6. $\frac{5x-5y}{12x} \cdot \frac{24x^2}{7y-7x}$

7. $\frac{35x}{3x-24} \div \frac{7x^2}{x^2-x-56}$

8. $\frac{x^2-16}{x^2-49} \div \frac{x+4}{x-7}$

9. $\frac{x^2-4x+16}{2x-6} \div \frac{x^3+64}{x^2+x-12}$

10. $\frac{6x^2+8x-30}{3x^2+13x-56} \cdot \frac{2x^2+17x+21}{24x^2-4x-60}$

11. $\dfrac{x^2-x-42}{x^3-3x^2-4x}\cdot\dfrac{x^2-5x-6}{x^2-1}\div\dfrac{x^2-36}{3x-12}$

12. $\dfrac{32x^2y^7}{x^2-y^2}\div\left[\dfrac{25x-125y}{x+y}\cdot\dfrac{48x^3y^5}{x^2-4xy-5y^2}\right]$

13. $\dfrac{x^2-49}{2x^2+9x-35}\div\dfrac{25x^3}{4x+40}\cdot\dfrac{50x^2}{x^2+3x-70}$

Determine the unknown expression in each equation.

14. $\left(\dfrac{x-5}{2x+3}\right)\cdot(?)=\dfrac{x^2-2x+3}{2x^2+x-3}$

15. $\left(\dfrac{6x^2-11x+3}{6x^2-13x-5}\right)\div(?)=\dfrac{3x-1}{3x+1}$

16. The number of items that a factory can produce in t hours per day is $N(t)=5t^2-5t$. The total cost of operating the factory for t hours is given by $C(t)=300t+300$.

(a) The average cost per item produced is given by $A(t)=C(t)\div N(t)$. Write a formula for the average cost per item.

(b) Use this function to complete the following table.

t	2	6	10	14	18	22
$A(t)$						

(c) Describe what happens to the average cost when the number of hours the factory operates is reduced to almost 1 hour.

(d) Describe what happens to the average cost when the number of hours the factory operates is increased to almost 24 hours.

9.3 Lecture Guide: Adding and Subtracting Rational Expressions

Objective 1: Add and Subtract rational expressions.

Adding or Subtracting Rational Expressions with the Same Denominator

Algebraically	Verbally	Algebraic Examples
If *A, B,* and *C* are real polynomials and $C \neq 0$, then $\frac{A}{C}+\frac{B}{C}=\frac{A+B}{C}$ and	To add two rational expressions with the same denominator, ____________ the numerators and use this common denominator.	$\frac{9}{28x}+\frac{5}{28x}=\frac{\qquad}{28x}$ $=\frac{\qquad}{28x}$ $=\_\_\_\_$
$\frac{A}{C}-\frac{B}{C}=\frac{A-B}{C}$	To subtract two rational expressions with the same denominator, ____________ the numerators and use this common denominator.	$\frac{9}{28x}-\frac{5}{28x}=\frac{\qquad}{28x}$ $=\frac{\qquad}{28x}$ $=\_\_\_\_$

Perform the indicated operations, and reduce the results to lowest terms. Assume the variables are restricted to values that prevent division by zero.

1. $\frac{2x}{4x^2y^3}+\frac{6x}{4x^2y^3}$

2. $\frac{6x+2y}{5x^3y^2}-\frac{x-8y}{5x^3y^2}$

3. $\frac{3x-8}{4x-12}-\frac{2x-5}{4x-12}$

4. $\frac{x^2+3x}{x^2-4}-\frac{6x+10}{x^2-4}$

5. $\frac{3x}{2x-1}-\frac{5}{1-2x}$

6. $\frac{2x}{x-3}+\frac{6}{3-x}$

Finding the LCD of Two or More Rational Expressions

Verbally	**Algebraic Example**
1. _______________ each denominator completely, including constant factors. Express repeated factors in exponential form.	Determine the LCD of $\frac{8}{10x^3y^2}$ and $\frac{5}{12x^2y^4}$ $10x^3y^2 =$ $12x^2y^4 =$
2. List each factor to the _______________ power to which it occurs in any single factorization.	$LCD=$
3. Form the LCD by _______________ the factors listed in Step 2.	$LCD=$

Determine the LCD of each pair of expressions.

7. $\frac{3x-1}{5x^2-20}$ and $\frac{3x}{7x+14}$

8. $\frac{2x+5}{4x^3-4x^2}$ and $\frac{5x-1}{3x^2-6x+3}$

Adding (Subtracting) Rational Expressions

Verbally	**Algebraic Example**
1. Express the denominator of each rational expression in factored form, and then find the LCD.	$\frac{7}{10x^3y^2}+\frac{5}{12x^2y^4}$
2. Convert each term to an equivalent rational expression whose _______________ is the LCD.	
3. Retaining the LCD as the denominator, add (subtract) the _______________ to form the sum (difference).	
4. Reduce the expression to lowest terms.	

Perform the indicated operations, and reduce the results to lowest terms. Assume the variables are restricted to values that prevent division by zero.

9. $\frac{1}{x-5}-\frac{2}{x+4}$

10. $\frac{3}{x+2}+\frac{1}{x-6}$

11. $\dfrac{2}{x^2+2x-35}-\dfrac{1}{3x^2-24x+45}$

12. $\dfrac{3}{x^2+4x-5}-\dfrac{1}{x^2-1}$

13. $\frac{-2x}{x-3}-\frac{3x}{3-x}-\frac{18}{x^2-9}$

14. $\frac{77}{x^2-x-30}+\frac{3}{x+5}-\frac{7}{x-6}$

15. Total Area of Two Regions Find the sum of the areas of the parallelogram and rectangle.

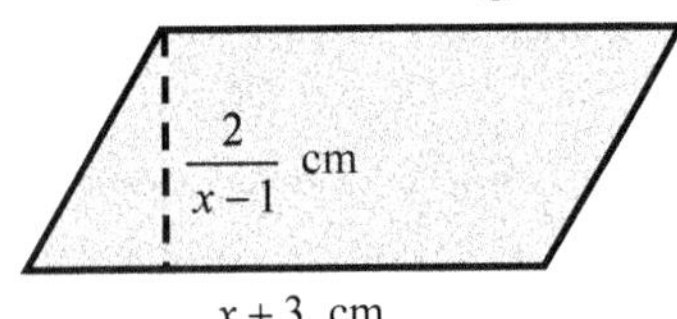

16. One person can load a truck in t hours while a second person would take $t+3$ hours. In 5 hours the fractional portion of the job done by each is $\frac{5}{t}$ and $\frac{5}{t+3}$, respectively.

(a) Write a rational function that gives the total amount of work completed by the two workers in 5 hours.

(b) Evaluate this function if $t=12$.

(c) Use the graph of this function to describe what happens to the total amount of work done as t becomes very large.

9.4 Lecture Guide: Combining Operations and Simplifying Complex Rational Expressions

Objective 1: Simplify rational expressions in which the order of operations must be determined.

Order of Operations

Step 1. Start with the expression within the ____________________ pair of grouping symbols.

Step 2. Perform all exponentiations.

Step 3. Perform all ____________________ and divisions as they appear from left to right.

Step 4. Perform all additions and ____________________ as they appear from left to right.

Use the correct order of operations to simplify each expression. Assume the variables are restricted to values that prevent division by zero.

1. $\dfrac{8x}{3}-\dfrac{4}{x^2}\cdot\dfrac{5x^3}{12}$

2. $\left(\dfrac{8x}{3}-\dfrac{4}{x^2}\right)\cdot\dfrac{5x^3}{12}$

3. $\left(\dfrac{3}{x}+\dfrac{1}{4}\right)\div\left(\dfrac{3}{x}-\dfrac{1}{4}\right)$

4. $\left(\dfrac{x^3}{x-3}\right)\cdot\left(\dfrac{1}{x}-\dfrac{3}{x^2}\right)^2$

Objective 2: Simplify complex fractions.

A **complex rational expression** is a rational expression where the numerator and/or the denominator also contain _______________. It is very important to identify the "main fraction bar" in a complex fraction.

Simplify:

5. $\dfrac{\frac{1}{2}}{5}$

6. $\dfrac{1}{\frac{2}{5}}$

Simplify by rewriting each expression with the division symbol $\div$. Assume the variables are restricted to values that prevent division by zero.

7. $\dfrac{\dfrac{x^2-9}{2x+6}}{4x}$

8. $\dfrac{\dfrac{x^2+2xy+y^2}{35x^2}}{\dfrac{x^2-y^2}{49x^3}}$

9. $\dfrac{\dfrac{3}{x}+\dfrac{5}{x^2}}{\dfrac{5x}{3}}$

10. $\dfrac{\dfrac{3x-24}{x^2-x-6}}{\dfrac{2}{x+2}-\dfrac{1}{x-3}}$

Simplify by multiplying both the numerator and the denominator by the LCD of all terms. Assume the variables are restricted to values that prevent division by zero.

11. $\dfrac{\dfrac{1}{3}-3}{\dfrac{5}{6}+\dfrac{1}{4}}$

12. $\dfrac{1+\dfrac{6}{x}+\dfrac{5}{x^2}}{1-\dfrac{25}{x^2}}$

Simplifying expressions containing negative exponents (Two Methods):

13. Simplify $\dfrac{1-2x^{-1}+x^{-2}}{x^{-1}-x^{-2}}$ by first converting each term with negative exponents to a term with positive exponents. Assume that x is restricted to values that prevent division by zero.

14. Simplify $\dfrac{1-2x^{-1}+x^{-2}}{x^{-1}-x^{-2}}$ by first multiplying the numerator and the denominator by the lowest power of x that will eliminate all of the negative exponents on x. Assume that x is restricted to values that prevent division by zero.

15. Simplify $\dfrac{x^{-1} - 9x^{-2} + 20x^{-3}}{1 - 5x^{-1}}$ assuming that x is restricted to values that prevent division by zero.

16. Write an expression that represents the area of this trapezoid.

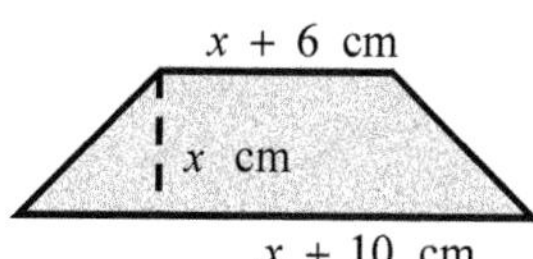

9.5 Lecture Guide: Solving Equations Containing Rational Expressions

Objective 1: Solve equations containing rational expressions.

Solving an Equation Containing Rational Expressions

Verbally	**Algebraic Example**
1. Multiply both sides of the equation by the __________.	$\dfrac{1}{x-2}+3=\dfrac{2x}{x-2}$
2. Solve the resulting equation.	
3. Check the solution to determine whether it is an excluded value and therefore ________________.	

Solve each equation.

1. $\dfrac{1}{x-3}-5=\dfrac{3x}{3-x}$

2. $\dfrac{x+6}{x+4}=\dfrac{2}{x+4}+5$

3. Solve: $\dfrac{5x-1}{x^2-2x-15}-\dfrac{3}{x-5}=\dfrac{2}{x+3}$

4. Solve: $\dfrac{5x-1}{x^2-2x-15}+\dfrac{3}{5-x}=\dfrac{2x}{x+3}$

5. Solve: $\dfrac{3}{x-2}+\dfrac{4}{x+3}=\dfrac{6x+3}{x^2+x-6}$

6. Simplify: $\dfrac{3}{x-2}+\dfrac{4}{x+3}-\dfrac{6x+3}{x^2+x-6}$

7. Solve: $\frac{2}{x} - \frac{x+8}{x+2} = \frac{12}{x^2+2x}$

8. Simplify: $\frac{2}{x} - \frac{x+8}{x+2} - \frac{12}{x^2+2x}$

Objective 2: Solve a rational equation for a specified variable.

Just as we did in Section 2.6, solving an equation for a specified variable involves isolating that variable on one side of the equation.

9. Solve $R = \frac{D}{T}$ for T.

10. Solve $x = \frac{yw}{z}$ for z.

11. Solve $\frac{1}{a} + \frac{1}{b} = \frac{1}{c}$ for a.

12. Solve $\frac{1}{a} + \frac{1}{b} = \frac{1}{c}$ for c.

13. Solve $I = \dfrac{E}{r_1 + r_2}$ for r_2.

14. Solve $\dfrac{1}{R} - \dfrac{1}{r_1} = \dfrac{1}{r_2}$ for R.

Determine the intercepts of each rational function. Use the given graph of each function to verify your results.

15. $f(x) = \dfrac{x+6}{x-2}$

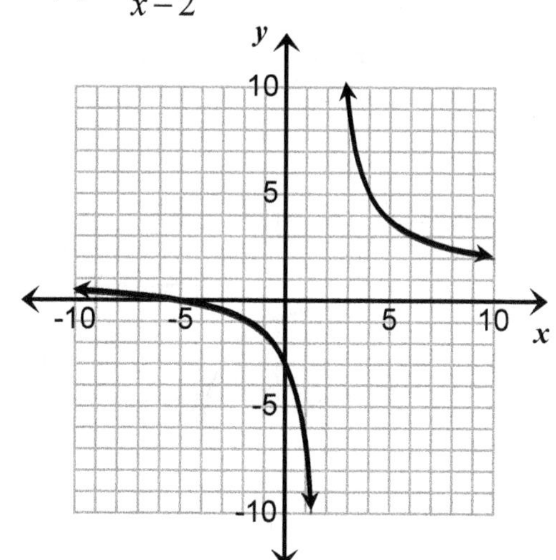

16. $f(x) = \dfrac{x^2 - 4}{x^2 - 5x - 6}$

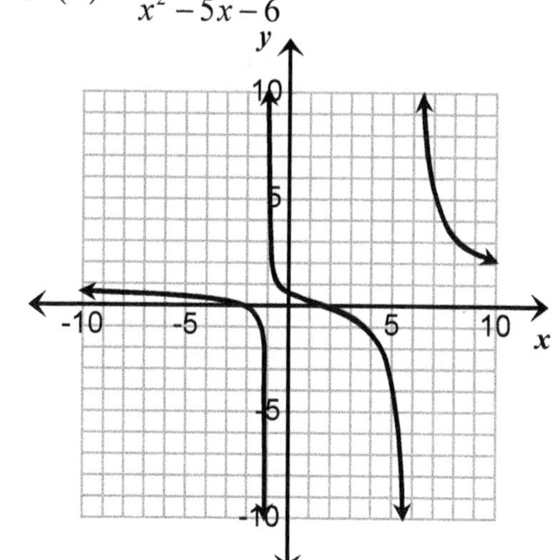

9.6 Lecture Guide: Inverse and Joint Variation and Other Applications Yielding Equations with Fractions

Objective 1: Distinguish between direct and inverse variation.

Comparison of Direct and Inverse Variation

If x and y are real variables, k is a real constant, and $k \neq 0$ then:

Verbally
y varies **directly** as x.

As the magnitude of x increases, the magnitude of y ________________ linearly.

Algebraically
$y = kx$

Example: $y = 2x$

Numerical Example

x	$y = 2x$
-3	-6
-1	-2
0	0
1	2
3	6

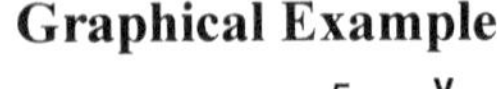

Graphical Example

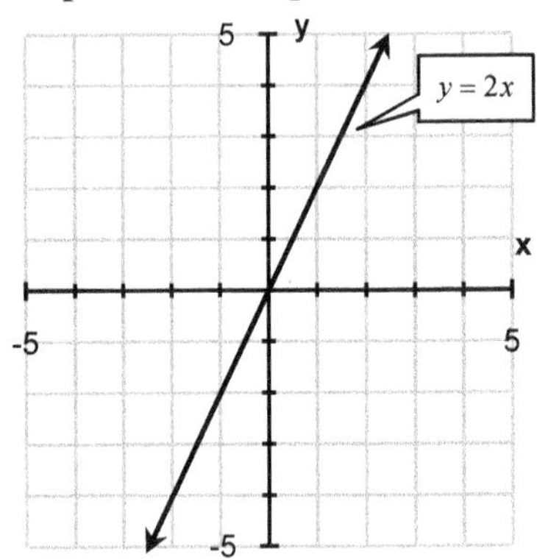

Verbally
y varies **inversely** as x.

As the magnitude of x increases, the magnitude of y ________________.

Algebraically
$y = \frac{k}{x}$ for $x \neq 0$

Example: $y = \frac{12}{x}$

Numerical Example

x	$y = \frac{12}{x}$
1	12
2	6
3	4
4	3
5	2.4

Graphical Example

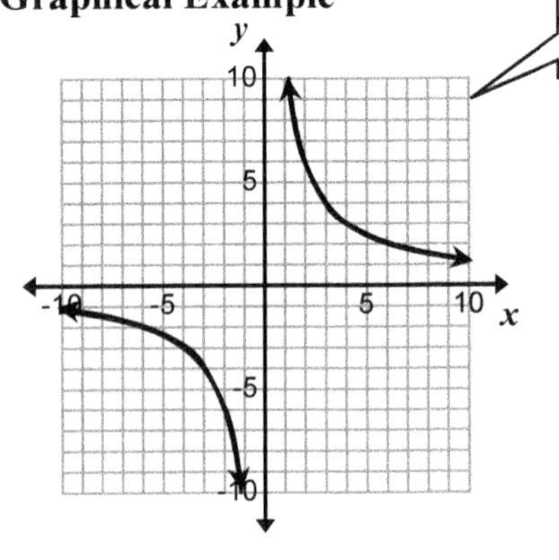

Joint Variation

If x, y, and z are real variables, k is a real constant, and $k \neq 0$ then:

Verbally	**Algebraically**
z varies jointly as x and y.	$z = kxy$

Determine whether the relationship illustrated is direct or inverse variation.

1. $p = 3t$

2. $a = \dfrac{12.5}{b}$

3.

x	y
1	120
2	60
3	40
4	30
5	26
6	20

4.

x	y
1	20
2	40
3	60
4	80
5	100
6	120

Objective 2: Translate statements of variation.

Write an equation for each statement of variation. Use k as the constant of variation.

5. P varies directly as T

6. P varies inversely as V

7. P varies directly as T and inversely as V

8. P varies jointly as T and V

Translate each equation into a verbal statement of variation.

9. $m = \dfrac{k}{n}$

10. $m = \dfrac{kn}{p}$

11. $m = knp$

Objective 3: Solve problems involving variation.

Use the given statement of variation to solve each problem.

12. y varies directly as x and $y = 45$ when $x = 9$. Find y when $x = 3$.

13. y varies inversely as x and $y = 5$ when $x = 9$. Find y when $x = 3$.

14. z varies directly as x and inversely as y and $z = 36$ when $x = \frac{1}{2}$ and $y = 8$.

Find z when $x = 5$ and $y = 12$.

15. z varies jointly as x and y and $z = 36$ when $x = \frac{1}{2}$ and $y = 8$. Find z when $x = 2$ and $y = 5$.

16. y varies directly as x. Use this information to complete this table.

x	y
2	24
4	
6	
8	
10	

17. y varies inversely as x. Use this information to complete this table.

x	y
2	24
4	
6	
8	
10	

18. The weight of an object on the moon varies directly as its weight on the earth. If an astronaut weighs 150 lbs on earth and 25 lbs on the moon.

(a) Determine the constant of variation.

(b) Write an equation relating the weight of an object on the moon to its weight on the earth.

(c) Use this equation to complete the table of values.

x, weight on the earth (lbs)	y, weight on the moon (lbs)
120	
130	
140	
150	
160	

19. The number of revolutions made by a wheel rolling a given distance varies inversely as the wheel's circumference. A wheel of circumference 20 cm makes 100 revolutions in going a certain distance.

(a) Determine the constant of variation.

(b) Write an equation relating the number of revolutions made by a wheel to its circumference for this given distance.

(c) Use this equation to complete the table of values.

c, circumference (cm)	n, number of revolutions (rev)
10	
15	
20	
25	
30	

Objective 4: Solve applied problems that yield equations with fractions.

Strategy for Solving Word Problems

Step 1. Read the problem carefully to determine what you are being asked to find.
Step 2. Select a ______________ to represent each unknown quantity. Specify precisely what each variable represents and note any restrictions on each variable.
Step 3. If necessary, make a ____________ and translate the problem into a word equation or a system of word equations. Then translate each word equation into an ______________ equation.
Step 4. Solve the equation or system of equations, and answer the question completely in the form of a sentence.
Step 5. Check the _____________________ of your answer.

20. The sum of the reciprocals of two consecutive odd integers is $\frac{8}{15}$. Find these integers.

(a) Identify the variable:

Let n = the smaller odd integer

Let _________________ = the larger odd integer

(b) Write the word equation:

The _________________ of the _________________ of the two integers is $\frac{8}{15}$.

(c) Translate the word equation into an algebraic equation:

(d) Solve this equation:

(e) Write a sentence that answers the question:

(f) Is this answer reasonable?

21. The ratio of two readings of a pressure gauge on a boiler is $\frac{9}{10}$. The first reading was 10 units below normal and the second was 10 units above normal. What is the normal reading?

22. Two boats having the same speed in still water depart simultaneously from a dock, traveling in opposite directions in a river that has a current of 6 miles per hour. After a period of time one boat is 54 miles downstream, and the other boat is 30 miles upstream. What is the speed of each boat in still water?

23. If one painter can paint a wall by himself in 8 hours and a second painter can paint the same wall by herself in 6 hours, how long will it take them to paint the wall when working together?

24. Two employees from the Roofing Company, can put new shingles on a house in 12 hours when they work together. It takes the inexperienced roofer 7 hours longer than the more experienced roofer to put new shingles on a house when working alone. How long would it take the more experienced roofer to put shingles on the house alone?

Lecture Guides for

Chapter 10

Square Root and Cube Root Functions and Rational Exponents

10.1 Lecture Guide: Evaluating Radical Expressions and Graphing Square Root and Cube Root Functions

Radical Notation

The number r is an nth root of x if $r^n = x$.

$\sqrt[n]{x}$ denotes the principal nth root of x

- x is the ________________.
- $\sqrt{\ }$ is the ________________ **symbol**, or the radical.
- The natural number n is the ________________ or the **order of the radical.**

Principal Square Root For any real number $x \geq 0$:

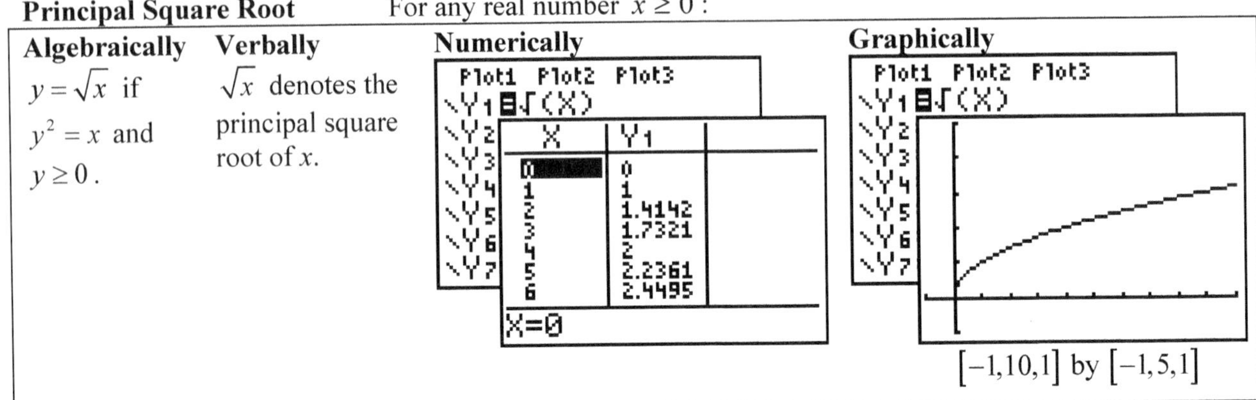

Principal Cube Root For any real number x:

Algebraically

$y = \sqrt[3]{x}$ if y is a real number and $y^3 = x$.

Verbally

$\sqrt[3]{x}$ denotes the principal cube root of x.

Numerically

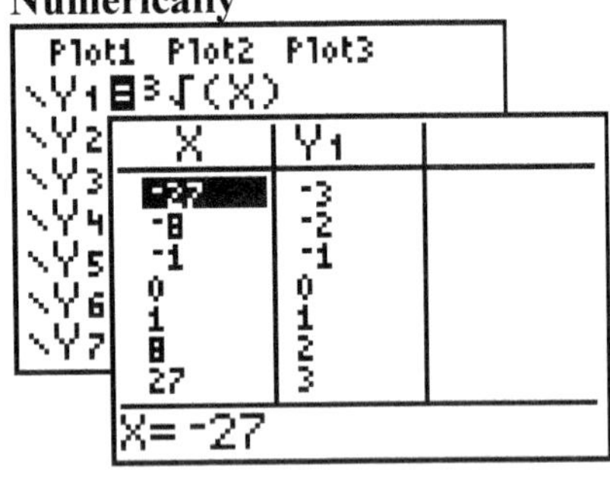

Graphically

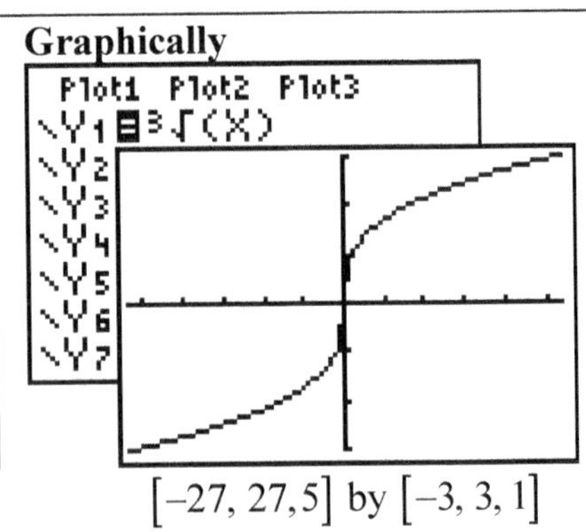

$[-27, 27, 5]$ by $[-3, 3, 1]$

Objective 1: Interpret and use radical notation.

Evaluate each expression. Use a calculator only to check your answer.

1. $\sqrt{81}$

2. $\sqrt[3]{27}$

3. $\sqrt[4]{81}$

4. $\sqrt[3]{-27}$

5. $\sqrt[4]{16}$

6. $\sqrt[3]{125}$

Evaluate each expression. Use a calculator only to check your answer.

7. $\sqrt{9}+\sqrt{16}$

8. $\sqrt{9+16}$

9. $\sqrt{\dfrac{9}{16}}$

10. $\sqrt[3]{\dfrac{-1}{27}}$

Objective 2: Graph and analyze square root and cube root functions.

11. $f(x)=\sqrt{x+4}$

Type:

***x*-intercept:**

***y*-intercept:**

Domain:

Range:

Table

x	y
–4	
–3	
0	
5	

Graph

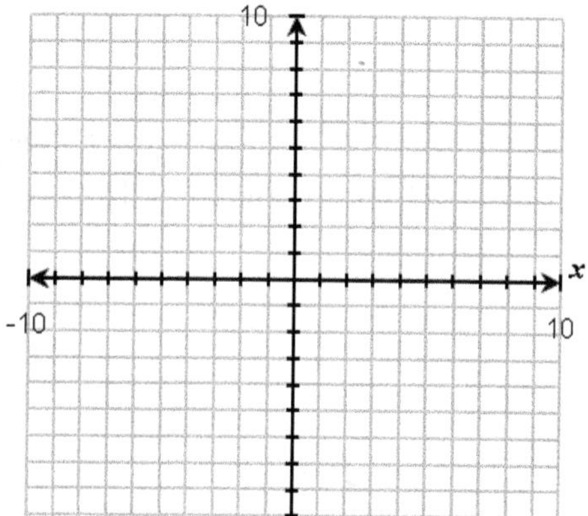

12. $f(x)=\sqrt[3]{x+1}$

Type:

***x*-intercept:**

***y*-intercept:**

Domain:

Range:

Table

x	y
–9	
–2	
–1	
0	
7	

Graph

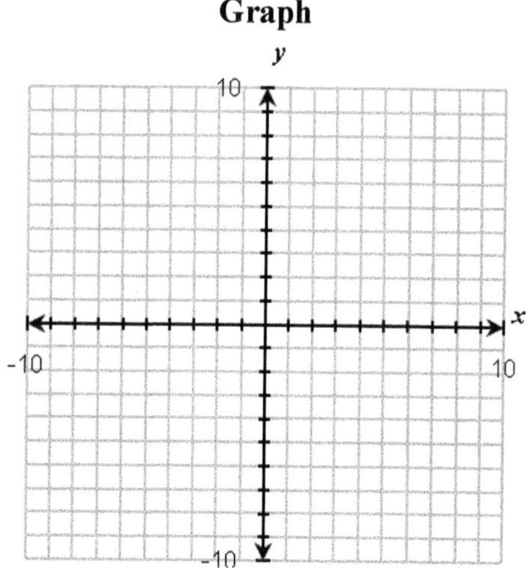

Domain of a Function $y = f(x)$

The domain of a real-valued function $y = f(x)$ is the set of all real numbers for which $f(x)$ is also a real number. This means the domain excludes all values that cause division by ________________ or cause a ________________ number under a square root symbol.

13. Determine the domain of $f(x) = \sqrt{2x+5}$ without using a calculator. Then graph $y_1 = \sqrt{2x+5}$ using a window of $[-5,5,1]$ by $[-5,5,1]$ to verify your result.

14. Determine the domain of $f(x) = \sqrt{5-2x}$ without using a calculator. Then graph $y_1 = \sqrt{5-2x}$ using a window of $[-5,5,1]$ by $[-5,5,1]$ to verify your result.

Determine the domain of the following. Use a graphing calculator only as a check on your work.

15. $f(x) = \sqrt[3]{4x+5}$

16. $f(x) = \dfrac{x}{x-3}$

17. $f(x) = \sqrt{4x+5}$

18. $f(x) = 2x^2 - 1$

19. $f(x) = 2x - 3$

20. $f(x) = |x-2| - 3$

Match each function with the appropriate graph. Hint: Complete these problems by recognizing the shape of the graph from the defining equation.

21. $f(x) = \sqrt[3]{4x+5}$ ______

22. $f(x) = 2x^2 - 1$ ______

23. $f(x) = \dfrac{x}{x-3}$ ______

24. $f(x) = |x-2| - 3$ ______

25. $f(x) = \sqrt{4x+5}$ ______

26. $f(x) = 2x - 3$ ______

A.

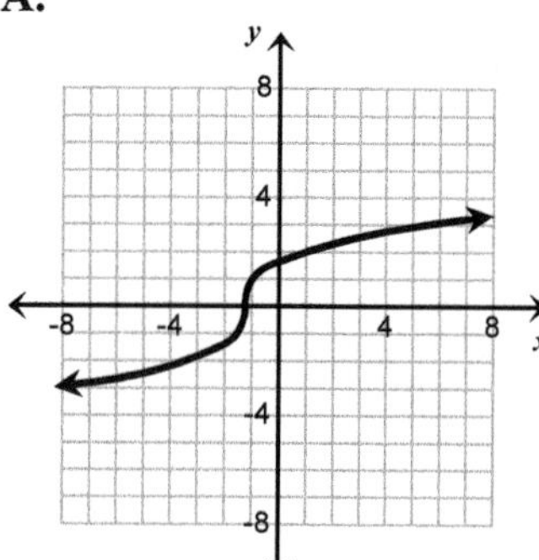

B.

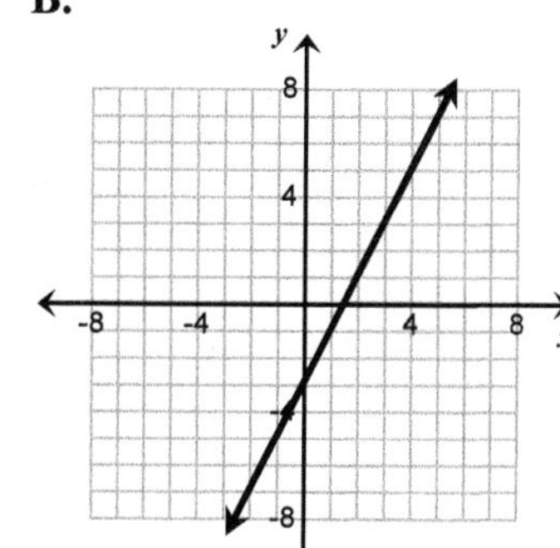

C.

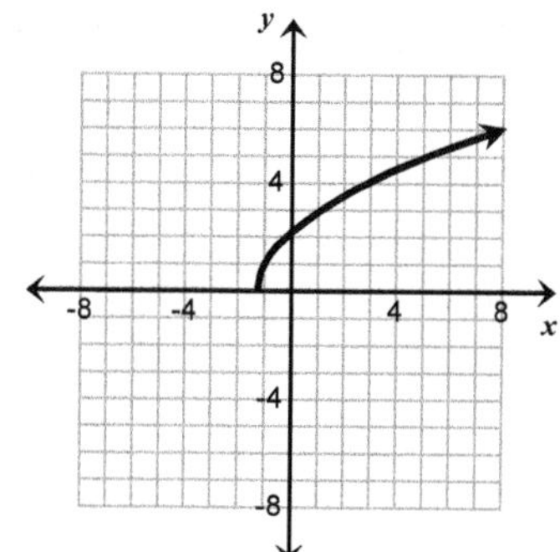

D.

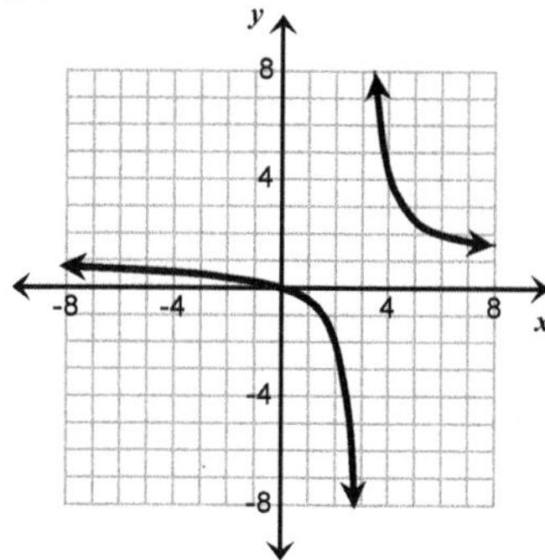

E.

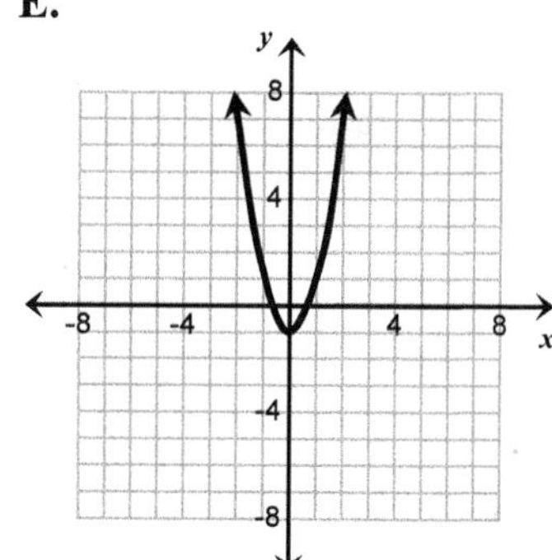

F.

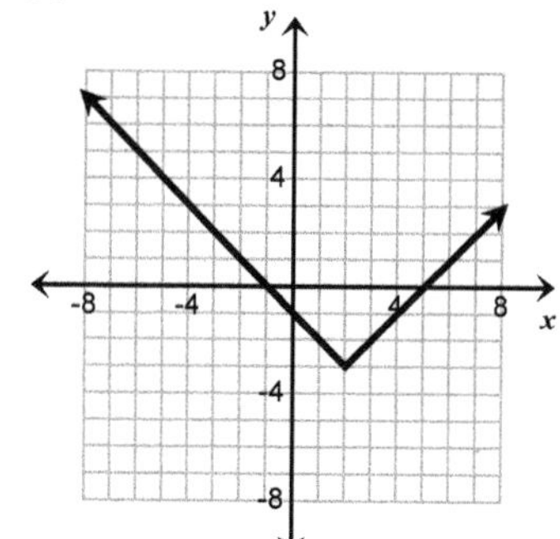

Mentally estimate the value of each expression and then use a calculator or a spreadsheet to approximate each value to the nearest thousandth.

27. $\sqrt{67}$

28. $\sqrt[3]{67}$

29. $\sqrt[4]{17}$

30. $\sqrt[3]{1+8}$

31. Determine to the nearest tenth of a centimeter the length of the hypotenuse of a right triangle with legs of 5 cm and 7 cm. Use the formula $c = \sqrt{a^2 + b^2}$.

32. Determine to the nearest tenth of a centimeter the radius of a sphere with a volume of 400 cm^3 . Use the formula $r = \sqrt[3]{\frac{3V}{4\pi}}$.

10.2 Lecture Guide: Adding and Subtracting Radical Expressions

Objective 1: Add and subtract radical expressions.

Like Radicals

We add like radicals in exactly the same way that we add like terms of a polynomial. Like radicals must have the same ______________ and the same ______________; only the ______________ of the like terms can differ. In exponential form like radicals have the same base and the same exponents.

Classify each pair of radicals as either like or unlike. (Circle the correct answer.)

1. $\frac{\sqrt{3}}{2}$ and $\frac{\sqrt{2}}{2}$ Like or Unlike?

2. $\sqrt[4]{5}$ and $-\sqrt[5]{5}$ Like or Unlike?

3. $\sqrt[3]{6}$ and $\sqrt[3]{7}$ Like or Unlike?

4. $3\sqrt[4]{6}$ and $5\sqrt[4]{6}$ Like or Unlike?

Find each sum or difference.

5. $\sqrt{100} - \sqrt{49}$

6. $5\sqrt{4} + 2\sqrt{9} - 4\sqrt{25}$

7. $2\sqrt[3]{5} + 9\sqrt[3]{5}$

8. $3\sqrt[5]{x} + 7\sqrt[5]{x} - 4\sqrt[5]{x}$

9. $6\sqrt[3]{5} - 9\sqrt[3]{5} + 3\sqrt[3]{5}$

10. $3x\sqrt{y} - 4x\sqrt{y} - 5x\sqrt{y}$

11. $5\sqrt[3]{5} - 6\sqrt{5} + 2\sqrt[3]{5} + 3\sqrt{5}$

12. $3\sqrt{xy} + 7\sqrt[3]{xy} - 4\sqrt{xy} - 9\sqrt[3]{xy}$

Objective 2: Simplify radical expressions.

Properties of Radicals

If $\sqrt[n]{x}$ and $\sqrt[n]{y}$ are both real numbers, then

Radical Notation	**Verbally**	**Radical Examples**
$\sqrt[n]{xy} = \sqrt[n]{x}\sqrt[n]{y}$	The nth root of a product equals the product of the nth roots.	$\sqrt{75} =$
$\sqrt[n]{\frac{x}{y}} = \frac{\sqrt[n]{x}}{\sqrt[n]{y}}$ for $y \neq 0$	The nth root of a quotient equals the quotient of the nth roots.	$\sqrt{\frac{16}{25}} =$

*If either $\sqrt[n]{x}$ or $\sqrt[n]{y}$ is not a real number, these properties do not hold true.

When simplifying radical expressions, it is very useful to recognize perfect n^{th} powers. The following table might help.

$1^2 =$	$2^2 =$	$3^2 =$	$4^2 =$	$5^2 =$	$6^2 =$
$7^2 =$	$8^2 =$	$9^2 =$	$10^2 =$	$11^2 =$	$12^2 =$
$1^3 =$	$2^3 =$	$3^3 =$	$4^3 =$	$5^3 =$	$6^3 =$
$1^4 =$	$2^4 =$	$3^4 =$	$4^4 =$	$1^5 =$	$2^5 =$

Simplify each radical expression.

13. $\sqrt{48}$

14. $\sqrt{108}$

15. $\sqrt[3]{48}$

16. $\sqrt[3]{108}$

17. $\sqrt[4]{162}$

18. $\sqrt[4]{32}$

Find the sum or difference. Assume that all variables represent positive real numbers so that absolute value notation is not necessary.

19. $\sqrt{45}+\sqrt{20}$

20. $4\sqrt[3]{108x^3}-5\sqrt[3]{32x^3}$

21. $\sqrt{\frac{3}{16}}+\sqrt{\frac{3}{4}}$

22. $6\sqrt{\frac{3x}{25}}-4\sqrt{\frac{3x}{49}}$

23. $\sqrt{28x^3}-3\sqrt{700x^3}$

24. $2\sqrt[3]{40}+3\sqrt[3]{135}$

25. $5\sqrt[3]{54x^2}-2\sqrt[3]{128x^2}$

26. $\sqrt{4x^2+24xy+36y^2}-\sqrt{x^2}+\sqrt{y^2}$

27. **(a)** Use $y_1 = \sqrt{x^2}$ to complete the table to the right.

x	y_1
-3	
-2	
-1	
0	
1	
2	
3	

After examining the table, decide whether you think that $\sqrt{x^2}$ is **always / sometimes / never** equal to x. Circle the correct choice.

Evaluate the following expressions:

28. **(a)** $\sqrt{(-3)^2}$ **(b)** $\sqrt{(3)^2}$ **29.** **(a)** $\sqrt[3]{(-2)^3}$ **(b)** $\sqrt[3]{(2)^3}$

30. **(a)** $\sqrt[4]{(-1)^4}$ **(b)** $\sqrt[4]{(1)^4}$ **31.** **(a)** $\sqrt[5]{(-5)^5}$ **(b)** $\sqrt[5]{(5)^5}$

Note that we must be careful when evaluating radical expressions with an even index. If there are variables under a radical, we must use caution in how we handle these variables because they might represent negative values.

We have observed that $\sqrt[n]{x^n}$ is not always equal to x. In fact $\sqrt[n]{x^n} = x$ if n is ____________________ and $\sqrt[n]{x^n} = |x|$ if n is ____________________.

The only time we have to consider using absolute value notation is when n is ____________________.

Simplify each radical expression where x represents any real number. Use absolute-value notation wherever necessary.

32. $\sqrt{49x^2}$ **33.** $\sqrt[3]{-8x^3}$

34. $\sqrt{x^{10}}$ **35.** $\sqrt{x^4}$

36. $\sqrt[5]{x^{10}}$ **37.** $\sqrt[3]{x^6}$

38. $\sqrt[3]{64x^3}$ **39.** $\sqrt[4]{x^{16}}$

Use a calculator to approximate each expression to the nearest thousandth.

40. $\sqrt{50}$

41. $\sqrt[3]{50}$

42. $\sqrt[4]{50}$

43. $\sqrt[5]{50}$

44. $\dfrac{1-\sqrt{50}}{2}$

45. $\dfrac{1+\sqrt{50}}{2}$

46. $\dfrac{-3-\sqrt{5.25}}{4}$

47. $\dfrac{-3+\sqrt{5.25}}{4}$

Simplify each expression. Use absolute value notation only when it is needed.

	For $x \geq 0$	**For Any Real Number x**
48. $\sqrt{9x^2}$		
49. $\sqrt[3]{64x^3}$		

Simplify each radical expression. Assume $x \geq 0$ and $y \geq 0$.

50. $\sqrt{75x^3}$

51. $\sqrt{32x^5y^6}$

52. $\sqrt[3]{250x^4}$

53. $\sqrt[3]{24x^5y^7}$

10.3 Lecture Guide: Multiplying and Dividing Radical Expressions

Objective 1: Multiply radical expressions.

Multiplying Radicals

To multiply and divide some radical expressions, we use the properties:

$\sqrt[n]{x}\sqrt[n]{y}=\sqrt[n]{xy}$ for $x\geq 0$ and $y\geq 0$ $\qquad$ $\dfrac{\sqrt[n]{x}}{\sqrt[n]{y}}=\sqrt[n]{\dfrac{x}{y}}$ for $x\geq 0$ and $y>0$

Perform each indicated multiplication, and then simplify the product. Use your calculator to evaluate the original expression and your answer as a check on your answer.

1. $\sqrt{5}\sqrt{10}$

2. $\sqrt{3}\left(2\sqrt{3}-1\right)$

Perform each indicated multiplication, and then simplify the product. Assume $x>0$ and $y>0$.

3. $\sqrt{32}\sqrt{2}$

4. $\sqrt{2x}\sqrt{8x}$

5. $\sqrt[3]{27x}\sqrt[3]{x^2}$

6. $\left(3\sqrt{5}\right)\left(2\sqrt{3}\right)$

7. $4\sqrt{5}\left(3\sqrt{5}+7\right)$

8. $3\sqrt{3}\left(2\sqrt{5}-5\sqrt{3}\right)$

9. $(5x+4)(5x-4)$

10. $\left(\sqrt{7}-5\right)\left(\sqrt{7}+5\right)$

11. $\left(8\sqrt{7}-1\right)\left(2\sqrt{7}+3\right)$

12. $\left(5\sqrt{y}-3\sqrt{x}\right)\left(5\sqrt{y}+3\sqrt{x}\right)$

Conjugates Radicals: The conjugate of $\sqrt{x}+\sqrt{y}$ is $\sqrt{x}-\sqrt{y}$.

Write the conjugate of each expression. Then multiply the expression by its conjugate. (HINT: Review the special product $(A+B)(A-B)$.)

Expression	Conjugate	Product
13. $3-\sqrt{7}$		
14. $x+3\sqrt{y}$		
15. $x-\sqrt{2x+1}$		

Objective 2: Divide and simplify radical expressions.

Dividing Radicals
When dividing expressions involving radicals, it is customary to simplify the result to a form that does not contain any radicals in the denominator.

Recall that $\sqrt{x^2}$ = ____________ for $x>0$. In general, $\sqrt[n]{x^n}$ = _____________ for $x>0$.

As a warm-up to rationalizing the denominator of a radical expression, perform each multiplication and simplify the result.

16. $\sqrt{2}\sqrt{2}$

17. $\sqrt{8}\sqrt{2}$

18. $\sqrt[3]{5}\,\sqrt[3]{5^2}$

19. $\sqrt[3]{2x^2}\,\sqrt[3]{4x}$

Perform each division, and then express the quotient in rationalized form. Assume $x>0$ and $y>0$.

20. $\dfrac{\sqrt{2}}{\sqrt{5}}$

21. $\dfrac{\sqrt{3}}{\sqrt{7}}$

22. $\dfrac{\sqrt{5x}}{\sqrt{2y}}$

23. $\dfrac{1}{\sqrt[3]{2}}$

Perform each division, and then express the quotient in rationalized form. Assume $x > 0$.

24. $\sqrt[3]{\frac{x}{4}}$

25. $\frac{4}{\sqrt[3]{2x^2}}$

Perform the indicated divisions by rationalizing the denominator and then simplifying. Assume that all variables represent positive real numbers.

26. $\frac{16}{\sqrt{7}+3}$

27. $\frac{8}{3-\sqrt{5}}$

28. $\frac{\sqrt{3}}{2+\sqrt{5}}$

29. $\frac{4x}{\sqrt{x}-\sqrt{y}}$

30. $\frac{3}{\sqrt{5x}+\sqrt{7y}}$

31. $\frac{\sqrt{x}}{\sqrt{5x}-\sqrt{2y}}$

10.4 Lecture Guide: Solving Equations Containing Radical Expressions

Objective 1: Solve equations containing radical expressions.

Power Theorem

For any real numbers x and y and natural number n:

Algebraically	**Verbally**	**Example**
If $x = y$, then $x^n = y^n$.	If two expressions are equal, then their nth powers are equal.	If $\sqrt{x} = 5$, then $\left(\sqrt{x}\right)^2 = (5)^2$ $x = 25$

Caution: The equations $x = y$ and $x^n = y^n$ are not always equivalent. The equation $x^n = y^n$ can have a solution that is not a solution of $x = y$.

Caution:
$(-5)^2 = 5^2$ but $-5 \neq 5$

Solving Radical Equations Containing a Single Radical

To solve equations containing radical expressions, we will use the power theorem, which states that if two expressions are equal, then their nth powers are equal. For example, if $x = y$ then $x^2 = y^2$.

Procedure	**Example**
Step 1. Isolate a radical term (of order n) on one side of the equation.	$\sqrt{x+5} + 3 = 8$
Step 2. Raise both sides to the nth power.	
Step 3. Solve the resulting equation.	
Step 4. Check the possible solutions in the original equation to determine whether they are really solutions or are extraneous.	

Solve each equation.

1. $\sqrt{x-3}-2=0$

2. $\sqrt{2x-8}+4=3$

3. $\sqrt[3]{3x+8}=5$

4. $\sqrt{2x+4}=\sqrt{8x+7}$

5. $\sqrt{x-3}=\sqrt{2x+12}$

6. $\sqrt{60x-36}=5x$

7. $\sqrt{x^2+x+3}-x=3$

8. $\sqrt{x+7}=x-5$

9. Use the table and graph to determine the solution of the equation $\sqrt{3x+1} = x - 1$.

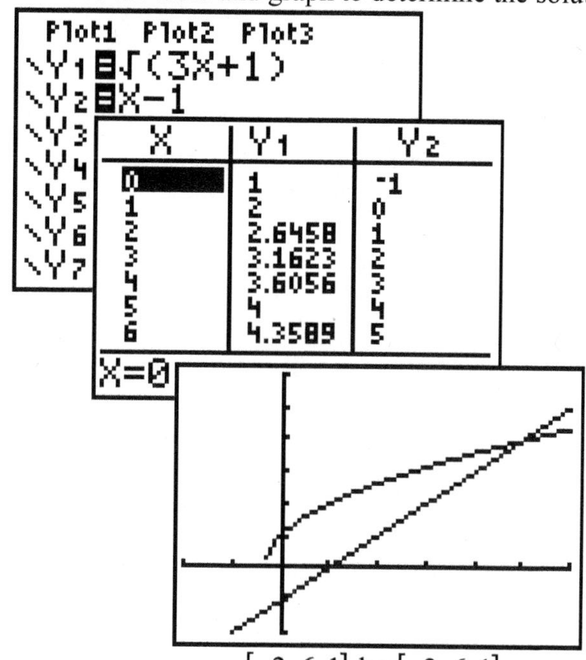

$[-2, 6, 1]$ by $[-2, 6, 1]$

Solution based on table and graph:

____

Solve this equation algebraically.

Determine the exact x- and y-intercepts of the graph of each function.

10. $y = \sqrt{x+4} - 7$

11. $y = \sqrt[3]{x-1} - 2$

The Pythagorean Theorem: If triangle ABC is a right triangle, then $a^2 + b^2 = c^2$

Use the Pythagorean Theorem to find the length of the side that is not given.

12.

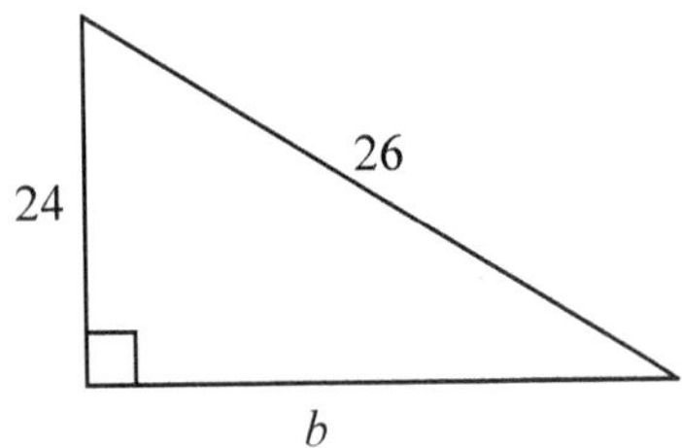

13.

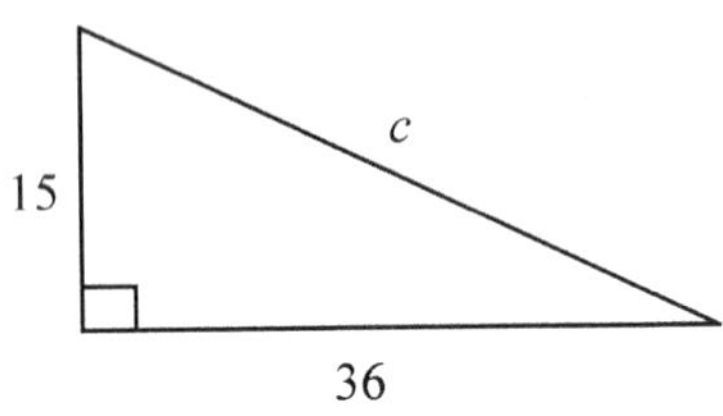

Objective 2: Calculate the distance between two points.

Distance Formula

The distance d between (x_1, y_1) and (x_2, y_2) is given by $d = \sqrt{(x_2 - x_1)^2 + (y_2 - y_1)^2}$.

Plot each pair of points and then calculate the distance between these points:

14. $(5, 1)$ and $(-3, 7)$

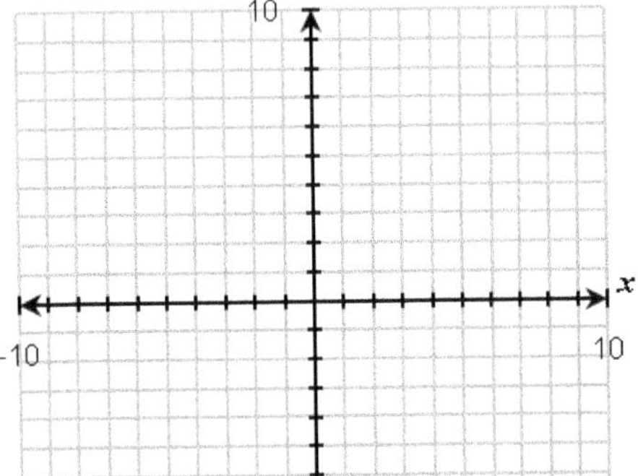

15. $(-1, -6)$ and $(-7, 9)$

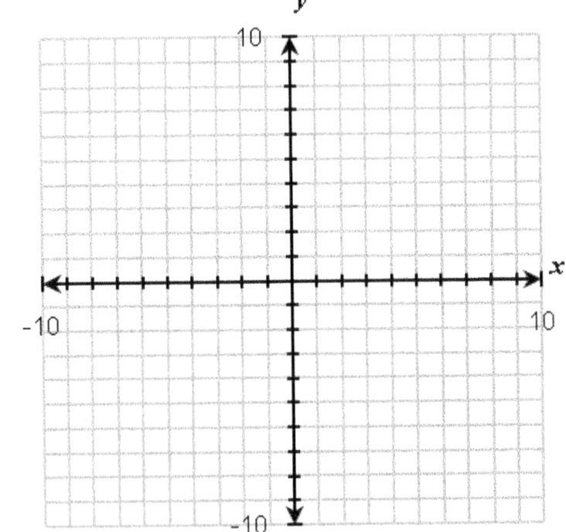

Calculate the distance between these points:

16. $(-14, -10)$ and $(-10, -3)$

17. $\left(\frac{1}{6}, \frac{1}{2}\right)$ and $\left(\frac{13}{6}, \frac{-3}{2}\right)$

18. Use the distance formula to determine the perimeter of the triangle formed by these points. Then use the Pythagorean theorem to determine whether the triangle formed is a right triangle. (Hint: Does $a^2 + b^2 = c^2$?)

$(-9, -5)$, $(-3, 4)$, and $(8, -3)$

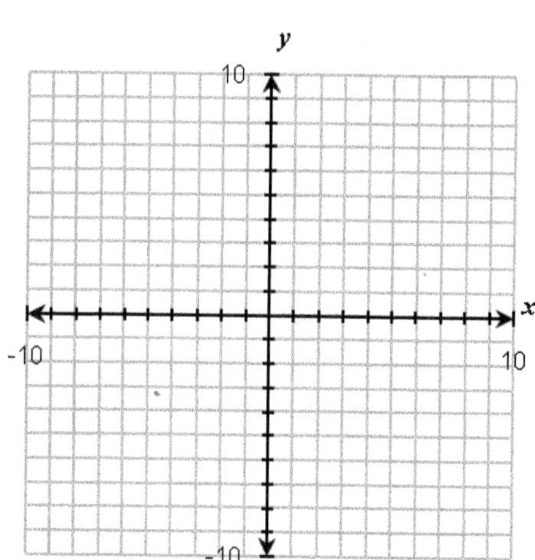

19. Find all points with an x-coordinate of 3 that are 10 units away from the point $(-5, 2)$.
Hint: Use the grid below to help.

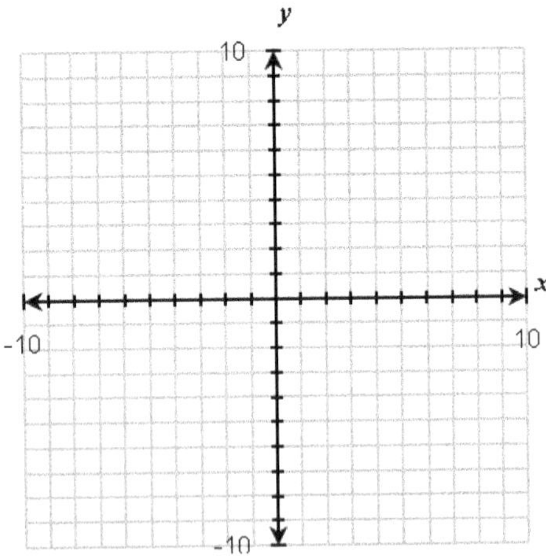

20. An extension cord is plugged into an outlet at point **A** on the side of a house. The owner of the house is running an electric weed trimmer to trim along a fence that runs parallel to the side of the house 40 ft away.

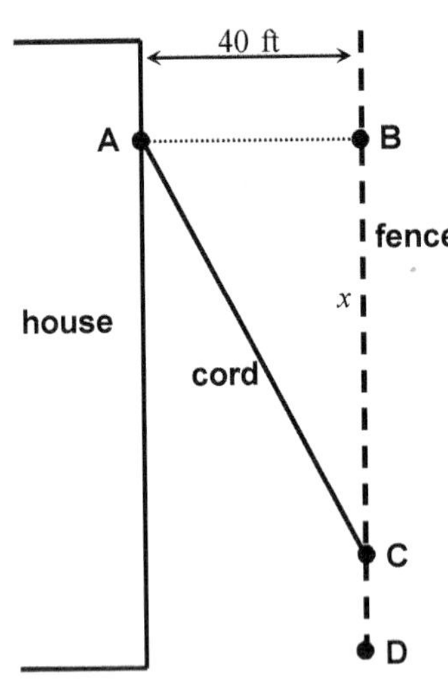

(a) Write a function $L(x)$ that gives the length of the extension cord needed to reach any point **C** along the fence, where x is the distance from **B** to **C** and the line segment from **A** to **B** is perpendicular to the side of the house.

(b) Evaluate and interpret $L(25)$.

(c) If point **D** is 32 ft from point **B**, will a 50 ft extension cord allow the owner to reach point **D?**

10.5 Lecture Guide: Rational Exponents and Radicals

Objective 1: Interpret and use rational exponents.

Principal n^{th} Root

The principal n^{th} root of the real number x is denoted by either $x^{1/n}$ or $\sqrt[n]{x}$.

	Verbally	**Examples in Radical Notation**	**Examples in Exponential Notation**
For $x > 0$:	The principal n^{th} root is positive for all natural numbers n.	$\sqrt{16} =$ $\sqrt[3]{27} =$	$16^{1/2} =$ $27^{1/3} =$
For $x = 0$:	The principal n^{th} root of 0 is 0.	$\sqrt{0} = 0$ $\sqrt[3]{0} = 0$	$0^{1/2} = 0$ $0^{1/3} = 0$
For $x < 0$:	If n is odd, the principal n^{th} root is negative.	$\sqrt[3]{-64} =$ $\sqrt[5]{-32} =$	$(-64)^{1/3} =$ $(-32)^{1/5} =$
	If n is even, there is no real n^{th} root. (The nth roots will be imaginary.)	$\sqrt{-1}$ is not a real number.	$(-1)^{1/2}$ is not a real number.

Use the product rule for exponents to simplify the following expressions:

1. $x^{1/2} \cdot x^{1/2}$

2. $x^{1/3} \cdot x^{1/3} \cdot x^{1/3}$

3. $x^{1/4} \cdot x^{1/4} \cdot x^{1/4} \cdot x^{1/4}$

$x^{1/2}$ is the principal ______________________ root of x.

$x^{1/3}$ is the principal ______________________ root of x.

$x^{1/4}$ is the principal ______________________ root of x.

Represent each expression by using exponential notation, and evaluate each expression.

4. $\sqrt{81}$

5. $\sqrt[4]{81}$

6. $\sqrt[7]{-1}$

7. $\sqrt[3]{-8}$

Represent each expression by using radical notation, and evaluate each expression.

8. $16^{1/2}$ **9.** $8^{1/3}$ **10.** $16^{1/4}$

11. $-49^{1/2}$ **12.** $(-49)^{1/2}$ **13.** $(-125)^{1/3}$

Rational Exponents

For a real number x and natural numbers m and n:

Algebraically	Examples in Radical Notation	Examples in Exponential Notation
If $x^{1/n}$ is a real number* $x^{m/n} = \left(x^{1/n}\right)^m = \left(\sqrt[n]{x}\right)^m$ or	$(-8)^{2/3} = \left(\sqrt[3]{-8}\right)^2$ $= (-2)^2$ $= 4$	$(-8)^{2/3} = \left[(-8)^{1/3}\right]^2$ $= (-2)^2$ $= 4$
$x^{m/n} = \left(x^m\right)^{1/n} = \sqrt[n]{x^m}$	$y^{\frac{3}{7}} = \sqrt[7]{y^3}$	$y^{\frac{3}{7}} = \left(y^3\right)^{\frac{1}{7}}$
$x^{-m/n} = \dfrac{1}{x^{m/n}}$, $x \neq 0$	$16^{-3/4} = \dfrac{1}{\left(\sqrt[4]{16}\right)^3}$ $= \dfrac{1}{2^3} = \dfrac{1}{8}$	$16^{-3/4} = \dfrac{1}{16^{3/4}}$ $= \dfrac{1}{\left(16^{1/4}\right)^3} = \dfrac{1}{2^3} = \dfrac{1}{8}$

*If $x < 0$ and n is even, then $x^{1/n}$ is not a real number.

Write each expression in radical notation and evaluate.

14. $16^{3/2}$ **15.** $8^{2/3}$

16. $(-8)^{2/3}$ **17.** $(81)^{3/4}$

18. $27^{-2/3}$ **19.** $(-125)^{-1/3}$

Objective 2: Use the properties of exponents.
You should be familiar with the properties of integer exponents from Chapter 5. Note that these properties apply to all rational exponents.

Properties of Exponents

Let m and n be real numbers and $x,\ x^m,\ x^n,\ y,\ y^m$, and y^n be nonzero real numbers.

Product rule: $x^m \cdot x^n = x^{m+n}$

Power rule: $\left(x^m\right)^n = x^{mn}$

Product to a Power: $(xy)^m = x^m y^m$

Quotient to a Power: $\left(\dfrac{x}{y}\right)^m = \dfrac{x^m}{y^m}$

Quotient rule: $\dfrac{x^m}{x^n} = x^{m-n}$

Negative power: $\left(\dfrac{x}{y}\right)^{-n} = \left(\dfrac{y}{x}\right)^n$

Simplify each expression. Assume that x is a positive real number.

20. $\left(5^{-6/7}\right)^{-7/2}$

21. $49^{1/8} \cdot 49^{3/8}$

22. $4^{1/3} \cdot 2^{1/3}$

23. $\dfrac{3^{13/5}}{3^{8/5}}$

24. $\left(\dfrac{16x^8}{81}\right)^{3/4}$

25. $\left(\dfrac{27x^9}{8}\right)^{-2/3}$

Simplify each expression. Assume that x and y are positive real numbers.

26. $\left(9^{1/9} \cdot 9^{2/9}\right)^{-9/2}$

27. $\dfrac{x^{1/3}}{x^{1/5}}$

28. $x^{2/5} \cdot x^{1/2}$

29. $\left(\dfrac{x^{2/3}}{x^{-5/3}}\right)^{6/7}$

30. $\left(125x^{9/4}y^{-3/2}\right)^{-2/3}$

31. $\dfrac{\left(xy^3\right)^{1/5}\left(x^2y\right)^{1/5}}{x^2y^{2/5}}$

Simplify each expression.

32. $2x^{3/5}\left(9x^{7/5} - x^{2/5} - 5x^{-3/5}\right)$

33. $\left(3x^{1/2} - 5y^{1/2}\right)\left(3x^{1/2} + 5y^{1/2}\right)$

34. $\left(3x + x^{-1/2}\right)^2$

35. $\left(a^{1/3} + 3\right)\left(a^{2/3} - 3a^{1/3} + 9\right)$

Use a graphing calculator or a spreadsheet to approximate each expression to the nearest hundredth.

36. $23^{\frac{1}{6}}$ **37.** $7^{\frac{2}{3}}$ **38.** $-3^{\frac{3}{4}}$

Lecture Guides for

Chapter 11

Exponential and Logarithmic Functions

11.1 Lecture Guide: Geometric Sequences and Graphs of Exponential Functions

Objective 1: Identify a geometric sequence.

Geometric Sequences

A **geometric sequence** is a sequence with a constant _______________ from term to term. That is, the ratio of any two consecutive terms is _______________. This constant is called the **common ratio** and is usually denoted by r.

Determine whether each sequence is geometric. If the sequence is geometric, find the common ratio r.

1. 2, 5, 8, 11, 14

2. 2, 5, 12.5, 31.25, 78.125

3. 10, 5, 2.5, 1.25, 0.625

4. 10, 5, 0, −5, −10

5. Write an arithmetic sequence of five terms with a first term $a_1 = 20$ and a common difference $d = -3$.

6. Write a geometric sequence of five terms with a first term $a_1 = 20$ and a common ratio $r = -3$.

7. Write a geometric sequence of five terms with a first term $a_1 = 20$ and a common ratio $r = \frac{1}{2}$.

Objective 2: Evaluate and graph exponential functions.

Exponential Function $f(x) = b^x$

Algebraically	**Verbally**	**Algebraic Example**	**Graphical Example**
If $b > 0$ and $b \neq 1$, then $f(x) = b^x$ is an exponential function with ______________ b.	An exponential function has a ______________ for the base and the variable is in the ______________.	$y = 2^x$	$y = 2^x$

8. Identify which functions are not exponential functions.

(a) $f(x) = 4^x$ **(b)** $f(x) = x^4$ **(c)** $f(x) = \left(\frac{1}{4}\right)^x$ **(d)** $f(x) = \left(-\frac{1}{4}\right)^x$ **(e)** $f(x) = 1^x$

Exponential Growth and Decay

Growth

If $b > 1$, then $f(x) = b^x$ exhibits exponential growth.

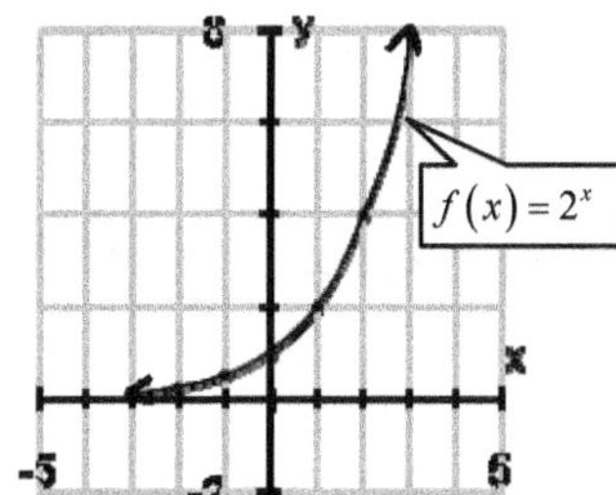

Decay

If $0 < b < 1$, then $f(x) = b^x$ exhibits exponential decay.

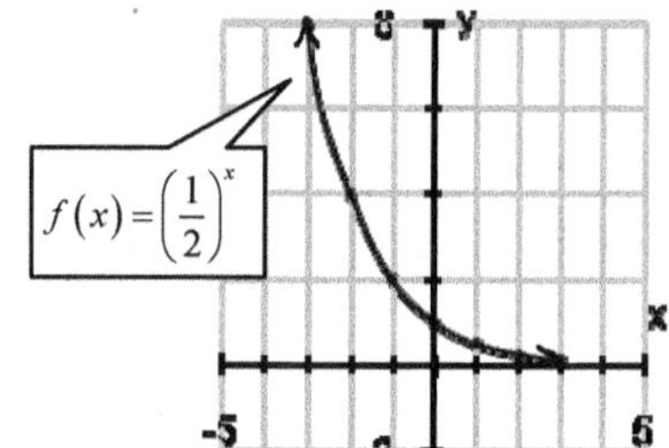

Properties of the Graphs of Exponential Functions

For the exponential function $f(x) = b^x$ with $b > 0$ and $b \neq 1$

1. The domain of f is the set of all real numbers, $\mathbb{R}$.
2. The range of f is the set of all positive real numbers.
3. There is no x-intercept.
4. The y-intercept is $(0,1)$.
5. **(a)** The growth function is asymptotic to the negative portion of the x-axis (it approaches but does not touch the x-axis).
   (b) The decay function is asymptotic to the positive portion of the x-axis.
6. **(a)** The growth function rises (or grows) from left to right.
   (b) The decay function falls (or decays) from left to right.

Complete each table and graph each function.

9. $f(x) = 3^x$

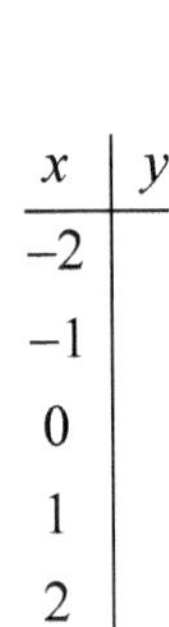

x	y
−2	
−1	
0	
1	
2	

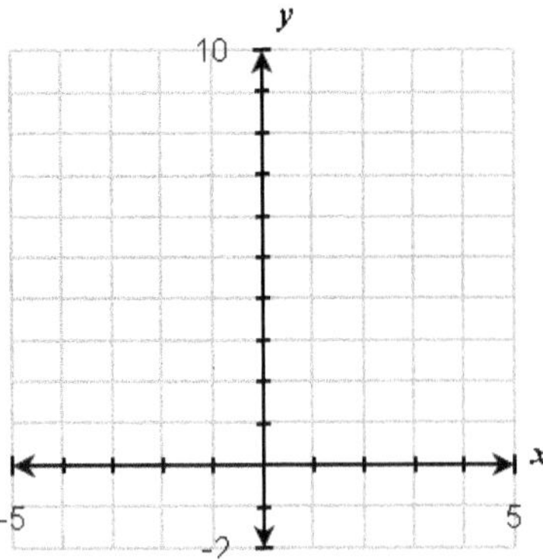

10. 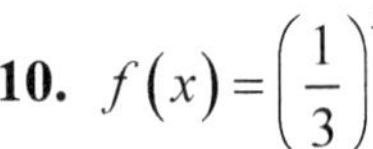
$f(x) = \left(\frac{1}{3}\right)^x$

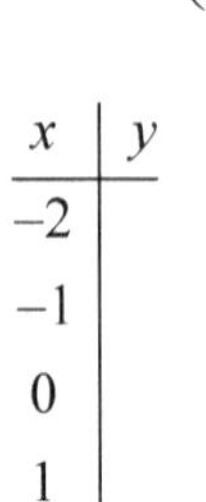

x	y
−2	
−1	
0	
1	
2	

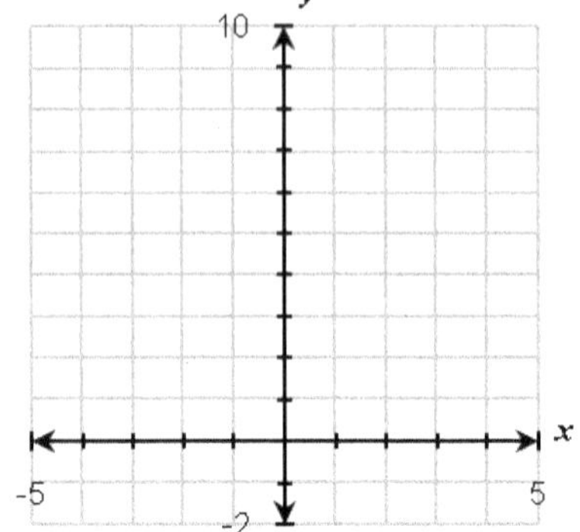

11. Given $f(x) = 2.4^x$ use a calculator or a spreadsheet to approximate each expression to the nearest thousandth.

(a) $f(6)$

(b) $f(-6)$

(c) $f(\pi)$

12. A share of stock on the NASDAQ is priced at \$100. Determine the price per share at the end of each of the next three years under each of the following assumptions.

(a) Arithmetic growth of \$12 per year.

(b) Arithmetic decay of \$12 per year.

(c) Geometric growth of 12% per year.

(d) Geometric decay of 12% per year.

13. The formula $A = P(1+r)^t$ can be used to compute the total amount of money A that accumulates when a principal P is invested at a yearly interest rate r and left to compound annually for t years. Use this growth formula to:

(a) Write a function for A for an investment of $7000 at 8.35% for t years.

(b) Evaluate this function for $t = 10$ years.

Properties of Exponential Equations

Algebraically	**Verbally**
For real exponents x and y and bases $a > 0, b > 0$,	
1. for $b \neq 1$, $b^x = b^y$ if and only if $x = y$.	**1.** The _______________ must be equal because the expressions are equal and have the same base.
2. for $x \neq 0$, $a^x = b^x$ if and only if $a = b$.	**2.** The _______________ must be equal because the expressions are equal and have the same exponents.

Objective 3: Solve simple exponential equations.

Solve each equation without using a calculator.

14. $2^x = 64$

15. $7^x = \frac{1}{7}$

16. $64^{8x} = 2$

17. $27^{3y} = 3$

18. $10^x = 0.0001$

19. $81^{5x+7} = 3$

20. $5^{3x-10} = 25$

21. $16^{6x+9} = 2$

22. $\left(\frac{3}{5}\right)^x = \frac{125}{27}$

23. $\left(\frac{3}{2}\right)^{-2x} = \frac{16}{81}$

24. $5^w = \sqrt[4]{5}$

25. $13^{3x+2} = \sqrt{13}$

11.2 Lecture Guide: Inverse Functions

Objective 1: Write the inverse of a function.

Inverse of a Function

If f is a function which matches each input value x with an output value y, then the ________________ of f reverses this correspondence to match this y-value with the x-value.

Complete the following table. Note in each case that f^{-1} undoes what f does.

Function f	Inverse f^{-1}
1. x → y 5 → 2 6 → 5 7 → 9	x → y ___ → ___ ___ → ___ ___ → ___
2. $\{(3,5),(-2,6),(0,4)\}$	
3. Loan amount | Monthly payment \$4000 | \$108.05 \$8000 | \$216.10 \$12000 | \$324.15 \$16000 | \$432.20	Monthly payment | Loan amount
4. $f(x)=x+5$; f increases x by 5.	
5. $f(x)=4x$; f multiplies x by 4.	
6. $f(x)=x^2$ for $x\geq 0$; f squares x	
7. $f(x)=\sqrt[4]{x}$; f takes the fourth root of x.	

Objective 2: Identify a one-to-one function.

One-to-One Function

A function is one-to-one if different input values produce ______________ output values.

If a function f is one-to-one, then f^{-1} will also be a function. We call f^{-1} an inverse function.

Determine whether each function is one-to-one.

8. $\{(-2,3),(2,5),(7,9),(8,3)\}$

9. $\{(-2,3),(2,5),(0,6),(1,4)\}$

The Horizontal Line Test

A graph of a function represents a one-to-one function if it is impossible to have any horizontal line ______________ the graph at more than one point.

10. Use the horizontal line test to determine whether each graph represents a one-to-one function.

(a)

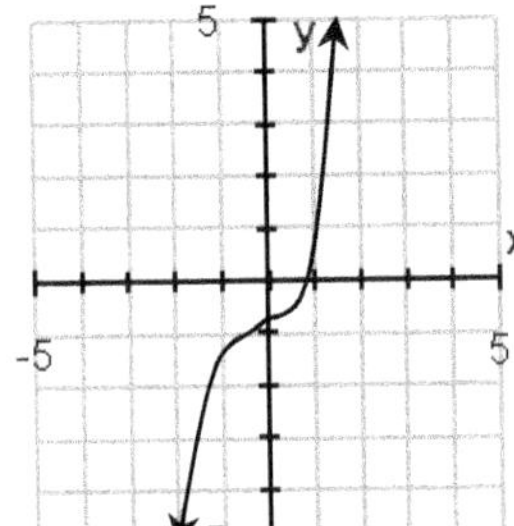

(b)

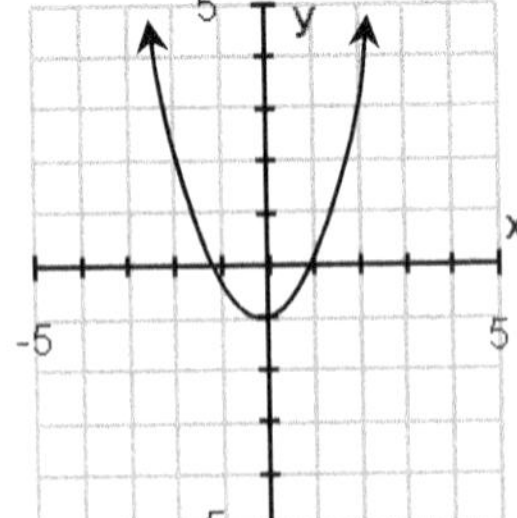

(c)

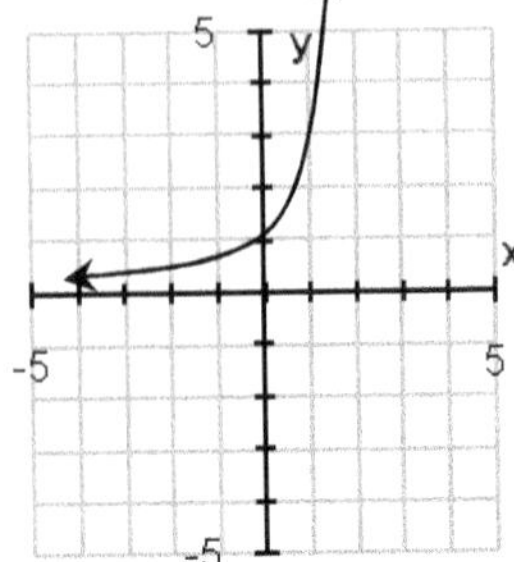

Finding an Equation for an Inverse Function

Verbally		**Algebraic Example**	
Step 1.	Replace $f(x)$ by *y*. Write this as a function of y in terms of x.	Function:	$f(x)=3x-5$
Step 2.	To form the inverse, replace each *x* with *y* and each *y* with *x*.	Inverse Function:	
Step 3.	If possible, solve the resulting equation for *y*.		
Step 4.	The inverse can then be written in function notation by replacing *y* with $f^{-1}(x)$.		

Find the inverse of each function.

11. $f(x)=\frac{1}{4}x+3$

12. $f(x)=3x-7$

13. $f(x)=\sqrt[3]{x-3}$

14. $f(x)=\sqrt[3]{x}-3$

Use $f(x) = 5x - 2$ and $f^{-1}(x) = \frac{x+2}{5}$ to evaluate each expression.

15. $f(3)$

16. $f^{-1}(13)$

17. $f(8)$

18. $f^{-1}(2)$

19. $f\left(\frac{1}{2}\right)$

20. $f^{-1}\left(\frac{1}{2}\right)$

Objective 3: Graph the inverse of a function.

The graph of a function and the graph of its inverse are symmetric about the line $x = y$.

21. Consider the graph to the right.

(a) Use the horizontal line test to determine if the function graphed to the right is a one-to-one function.

(b) Using the three ordered pairs graphed on the function to the right, graph the inverse of this function.

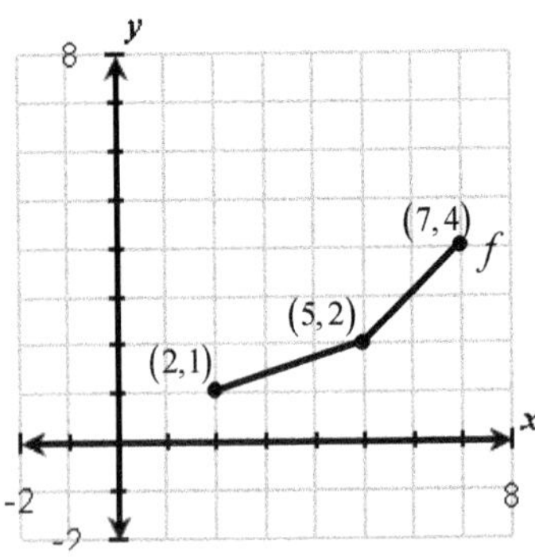

11.3 Lecture Guide: Logarithmic Functions

Objective 1: Interpret and use logarithmic notation.

For every logarithmic form there is a corresponding exponential form. Both forms contain the same information but from different perspectives.

Logarithmic Function, $f(x) = \log_b x$

Algebraically	**Verbally**	**Numerical Examples**
For $x > 0$, $b > 0$ and $b \neq 1$, $y = \log_b x$ if and only if $b^y = x$.	$y = \log_b x$ is read "y equals the log base b of x."	$\log_2 8 = 3$ because $2^3 = 8$ $\log_3 \frac{1}{9} = -2$ because $3^{-2} = \frac{1}{9}$

1. Complete the following table:

Logarithmic Form	**Verbal Description**	**Exponential Form**
$\log_5 25 = 2$		
$\log_2 32 = 5$		
$\log_7 \sqrt[3]{7} = 1/3$		
$\log_5 \sqrt[3]{25} = 2/3$		
$\log_2 \frac{1}{2} = -1$		
$\log_3 \frac{1}{27} = -3$		
	The log base 32 of 2 is $\frac{1}{5}$	
	The log base 9 of $\frac{1}{81}$ is -2.	
		$16^{3/4} = 8$
		$9^{3/2} = 27$
$\log_m p = q$		
		$a^b = c$

Objective 2: Evaluate simple logarithmic expressions.

Properties of Logarithmic Functions

For $b > 0$ and $b \neq 1$,

Logarithmic Form	Exponential Form	Numerical Examples
$\log_b 1 = 0$	$b^0 = 1$	$\log_3 1 =$
$\log_b b = 1$	$b^1 = b$	$\log_2 2 =$
$\log_b \frac{1}{b} = -1$	$b^{-1} = \frac{1}{b}$	$\log_7 \frac{1}{7} =$
$\log_b b^x = x$	$b^x = b^x$	$\log_3 3^4 =$

(These properties should be practiced until you are thoroughly comfortable with them.)

Determine the value of each logarithm by inspection.

2. $\log_3 81$

3. $\log_3 \frac{1}{27}$

4. $\log_{16} 4$

5. $\log_{2/5} \frac{5}{2}$

6. $\log_8 2$

7. $\log_3 0$

8. $\log_8 8^{2.7}$

9. $\log_2 (-4)$

10. $\log_{3/7} \frac{9}{49}$

11. $\log_5 (1)$

Solve each equation for x. Hint: Rewrite in exponential form if necessary.

12. $\log_5 25 = x$

13. $\log_9 27 = x$

14. $\log_x 16 = -4$

15. $\log_2 x = -1$

16. $\log_x \left(\frac{1}{25}\right) = -2$

17. $\log_8 x = \frac{2}{3}$

18. $\log_x \left(\frac{1}{243}\right) = -5$

19. $\log_2 (x+1) = 3$

Objective 3: Sketch the graph of a logarithmic function.

Comparison of Exponential Growth to Logarithmic Growth

Exponential Growth

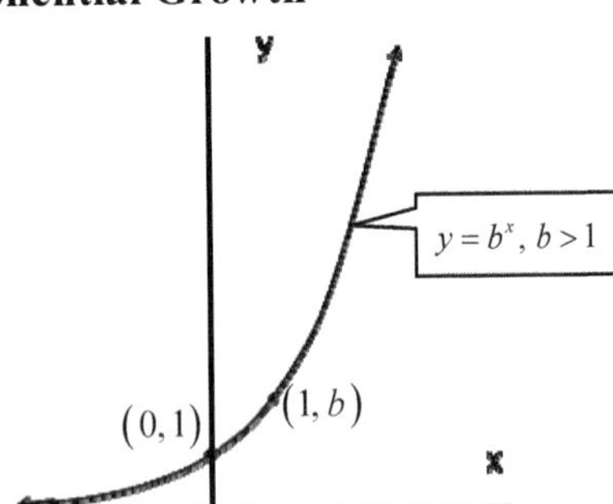

Features: For large values of x, extremely rapid growth.

x-intercept: None

y-intercept: $(0,1)$

Key point: $(1,b)$

Asymptotic to negative x-axis

Logarithmic Growth

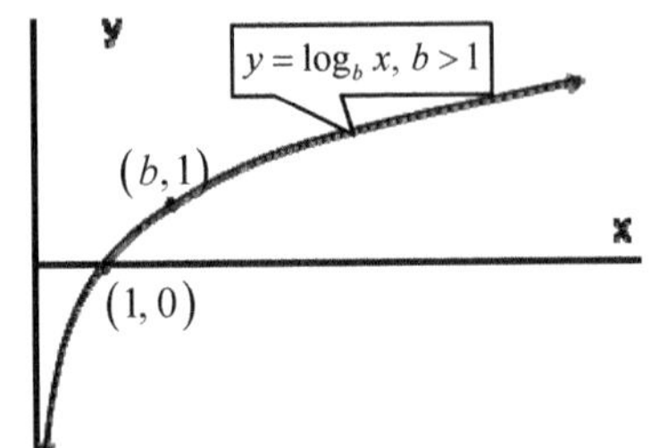

Features: For large values of x, extremely slow growth.

x-intercept: _________

y-intercept: _________

Key point: _________

Asymptotic to ____________________

20. Answer each question about exponential and logarithmic functions.

(a) Graph $f(x) = 2^x$ by completing the table and plotting these points

(b) Then use the concept of an inverse function to graph $f^{-1}(x)$ on the same grid.

x	$f(x) = 2^x$
-2	
-1	
0	
1	
2	
3	

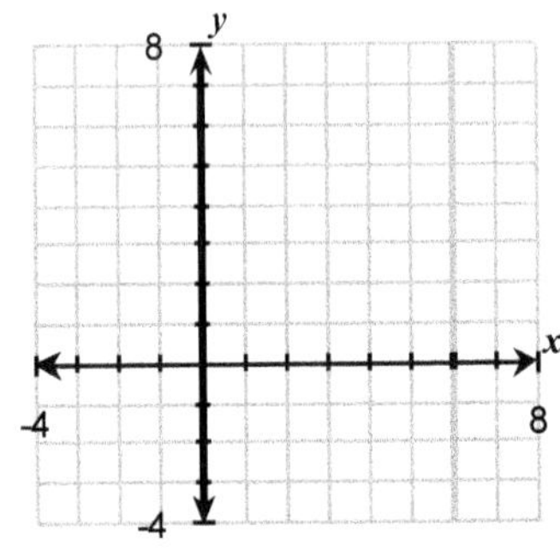

(c) On the second grid, graph $y = \log_2 x$ by completing the table and plotting these points.

(d) Compare the graph in part **d** to the graph in part **b**.

(e) The conclusion from part **e** is that the __________________ of $f(x) = 2^x$ is $f^{-1}(x) = \log_2 x$.

x	$y = \log_2 x$
$\frac{1}{4}$	
$\frac{1}{2}$	
1	
2	
4	
8	

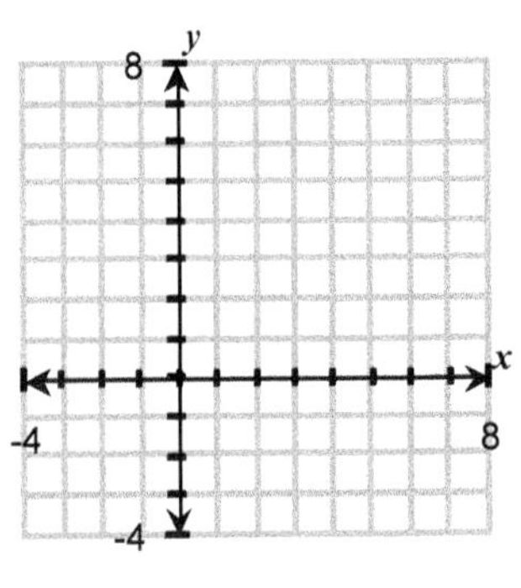

It is important to understand the relationship between exponential and logarithmic functions. If $f(x)=b^x$ with $b>0$ and $b\neq 0$ then for $x>0$ $f^{-1}(x)=\log_b x$. The function $f(x)=2^x$ has as its inverse the function $f^{-1}(x)=\log_2 x$. Notice: $3 \xrightarrow{2^x} 8$ and $8 \xrightarrow{\log_2 x} 3$. Thus, $(3, 8)$ is a point on the graph of the function $f(x)=2^x$ and $(8, 3)$ is a point on the graph of the inverse function $f^{-1}(x)=\log_2 x$

21. The domain of $f(x)=2^x$ is ________________, and the range of $f(x)=2^x$ is ________________. Since this graph passes the horizontal line test, it is ________________, and its inverse, $f^{-1}(x)=\log_2 x$, is a function. $f^{-1}(x)=\log_2 x$ has a domain of ________________ and a range of ________________.

11.4 Lecture Guide: Evaluating Logarithms

Objective 1: Evaluate common and natural logarithms with a calculator or a spreadsheet.

Common and Natural Logarithms

Common Logarithms	**Verbally**	**Numerical Example**
$\log x$ means $\log_{10} x$	$\log x$ is read "the (common) log of x" or "the log base 10 of x."	$\log 100 = 3$ because $10^3 = 1000$
Natural Logarithms		
$\ln x$ means $\log_e x$	$\ln x$ is read "the natural log of x" or "the log base e of x."	$\ln 100 \approx 4.605$ because $e^{4.605} \approx 100$

Use your calculator to approximate to the nearest hundredth the value of each logarithmic expression. See Technology Perspective 11.4.1.

1. $\log 105$

2. $\log 0.0125$

3. $\ln 6$

4. $\ln 0.5$

5. $\log 345{,}000{,}000{,}000$

6. $\ln 0.000000000123$

7. $\log\left(2.5^2\right)$

8. $\left(\ln 0.5\right)^2$

9. $\dfrac{\log 5}{6}$

10. $\dfrac{\log 5}{\log 6}$

11. $\log\left(\dfrac{5}{6}\right)$

12. $(\log 5)(\log 6)$

Solve each logarithmic equation. Hint: Convert from logarithmic form to exponential form.

13. $\log x = -1.699$

14. $\ln x = 3.2189$

11.5 Lecture Guide: Properties of Logarithms

Objective 1: Use the product rule, the quotient rule, and the power rule for logarithms.

Properties of Logarithms

For $x,\ y > 0$, $b > 0$, and $b \neq 1$,

	Algebraically	**Verbally**	**Numerical Example**
Product Rule	$\log_b xy = \log_b x + \log_b y$	The log of a product is the ______________ of the logs.	$\log_2 64 = \log_2 (4 \cdot 16) =$
Quotient Rule	$\log_b \frac{x}{y} = \log_b x - \log_b y$	The log of a quotient is the _______________ of the logs.	$\log_2 4 = \log_2 \frac{64}{16} =$
Power Rule	$\log_b x^p = p \log_b x$	The log of the pth power of x is p _____________ the log of x.	$\log_2 64 = \log_2 8^2 =$

Use one of the rules for logarithms to rewrite each expression and then use a calculator or a spreadsheet to verify that your new expression is equivalent to the original expression.

1. $\log(12 \cdot 5)$

2. $\ln\left(\frac{12}{5}\right)$

3. $\log 12^5$

Use the properties of logarithms to write each expression in terms of logarithms of simpler expressions. Assume that the argument of each logarithm is a positive real number.

4. $\log 2x$

5. $\log_3 x^3 y^2$

6. $\log_b \left(\frac{x-2}{y^3}\right)$

7. $\ln \sqrt[3]{xy^2}$

Combine these logarithms into a single logarithmic expression with a coefficient of 1. Assume that the argument of each logarithm is a positive real number.

8. $\log(x+1)+\log(x-3)$

9. $\log 2-2\log(x-4)$

10. $3\log x+2\log(x-1)$

11. $5\log x-\frac{1}{2}\log y$

Objective 2: Use the change-of-base formula to evaluate logarithms.

Change of Base Formula

For $a, b>0,\ a\neq 1$, and $b\neq 1$

For Logarithms	**Examples**
$\log_a x=\dfrac{\log_b x}{\log_b a}$ for $x>0$	$\log_3 x=$
For Exponents	
$a^x=b^{x\log_b a}$	$3^x=$

Use the change of base formula to evaluate:

12. $\log_2 17$

13. $\log_7 50$

14. $\log_7 33.4$

15. $\log_{0.52} 4.89$

16. Use the change-of-base formula for exponents to write $f(x) = 4^x$ as an exponential function with base e. Use a calculator or a spreadsheet to generate a table of values to check that the two functions are equivalent.

A Special Case Rule for Logarithms

For any positive real number x, $b > 0$, and $b \neq 1$:

Algebraically	**Numerical Example**
$b^{\log_b x} = x$	$12^{\log_{12} 51} =$

Determine the value of each expression. Then use the change of base formula and your calculator to check your results where possible.

17. $2^{\log_2 3}$

18. $x^{\log_x 5}$

19. $22^{\log_{22} 30}$

20. $10^{\log 0.0054}$

11.6 Lecture Guide: Solving Exponential and Logarithmic Equations

Objective 1: Solve exponential and logarithmic equations.

Solving Exponential Equations Algebraically

Verbally	Algebraic Examples
Step 1. **(a)** If it is obvious that both sides of the equation are powers of the same base, express each side of the equation as a power of this base and then equate the exponents.	**(a)** $2^x = 16$
(b) Otherwise, take the logarithm of both sides of the equation and use the power rule to form an equation that does not contain variable exponents.	**(b)** $2^x = 15$
Step 2. Solve the equation formed in Step 1.	

Solve each equation, and then use a calculator to approximate the solution to the nearest thousandth.

1. $3^{2x} = 81$

2. $5^{x+2} = 25^x$

3. $3^x = 7$

4. $3^{2x-1} = 5^x$

5. $e^{0.75t} = 4$

6. $(1.025)^{4t} = 10$

Solving Logarithmic Equations Algebraically

Verbally	Algebraic Examples
Step 1. **(a)** If possible, rewrite the logarithmic equation in exponential form.	**(a)** $\log(x+450)=3$ Check:
(b) Otherwise, use the properties of logarithms to write each side of the equation as a single logarithmic term with the same base. Then form a new equation by equating the arguments of these logarithms.	**(b)** $\ln 5x = \ln(3x+8)$
Step 2. Solve the equation formed in Step 1.	
Step 3. Check all possible solutions for extraneous values; logarithms of zero or negative arguments are undefined.	Check:

Solve each equation.

7. $\log_5(2x-1)=0$

8. $\log(4x+5)=1$

Solve each equation.

9. $\log(5x) = 2$

10. $\ln x + \ln(x-6) = \ln 7$

11. $\log_6 x + \log_6(x-1) = 1$

12. $\log(x^2 - 100) - \log(x+10) = \log 10$

11.7 Lecture Guide: Exponential Curve Fitting and Other Applications of Exponential and Logarithmic Equations

Objective 1: Use exponential and logarithmic equations to solve applied problems.

Growth and Decay Formulas

Periodic Growth Formula	**Verbally**	**Algebraic Example**
$A = P\left(1+\frac{r}{n}\right)^{nt}$	The amount A that is produced by an original amount P growing at an annual rate r with periodic compounding n times a year for t years.	$A = 1000\left(1+\frac{0.07}{4}\right)^{(4)(10)}$. This is the amount resulting from investing \$1000 at 7% compounded quarterly for 10 years.
Continuous Growth and Decay Formula	**Verbally**	**Algebraic Example**
$A = Pe^{rt}$	The amount A that is produced by an original amount P growing (decaying) continuously at a rate r for a time t. If $r > 0$ there is growth. If $r < 0$ there is decay.	$A = 1000e^{(0.07)(10)}$ This is the amount resulting from investing \$1000 at 7% compounded continuously for 10 years.

1. Compute the value of \$10,000 invested at 8% for 5 years if interest is compounded:

   (a) Semiannually

   (b) Quarterly

   (c) Weekly

   (d) Daily

   (e) Continuously

2. How many years will it take an investment of \$500 to triple in value if interest on the investment is compounded monthly at a rate of 9%?

3. Since the population of Springfield was last given to be 35,000 residents in 2000, it has been growing at a rate of 2.75% per year.

(a) Give a function to model the population of Springfield since 2000.

(b) Use this function to estimate the population in 2015.

(c) Using this growth rate determine when the population will be 60,000 residents.

4. A swab of a kitchen counter revealed 8,000 bacteria. This bacteria sample was found to grow at a continuous rate of 18% per hour when placed in an incubator. If the bacteria sample was placed in an incubator for 24 hours, how many bacteria will be present at the end of the period?

5. Use the formula $R = \log\left(\frac{A}{a}\right)$ to determine the magnitude of an earthquake on the Richter scale that measured 50,000 times the reference amplitude a.

6. The intensity of the music at a concert was measured at 9.9×10^{-6} watts/cm^2. Use the formula $D = 10 \log \frac{I}{I_o}$ to determine the decibel reading at this concert. $\left(I_o = 10^{-16} \text{ watts/cm}^2\right)$

7. Use the formula $\text{pH} = -\log \text{H}^+$ to determine H^+, the concentration of hydrogen ions in moles per liter of orange juice given that the pH for orange juice is 3.8.

8. Use a graphing calculator and the points given in the table to

(a) Draw a scatter diagram of these points.

(b) Determine the exponential function that best fits these points. Round the coefficients to three significant digits.

x	y
1	6.0
2	7.2
3	8.7
4	10.4
5	12.5
6	14.9

(c) Does this function represent exponential growth or exponential decay?

(d) Use this equation to estimate y when x is 2.5.

Lecture Guides for

Chapter 12

A Preview of College Algebra

12.1 Lecture Guide: Solving Systems of Linear Equations by Using Augmented Matrices

Objective 1: Use augmented matrices to solve systems of two linear equations with two variables.

A **matrix** is a ________________ array of numbers arranged into ______________ and ______________. Each row consists of entries arranged horizontally. Each column consists of entries arranged vertically. The ________________ of a matrix is given by stating first the number of rows and then the number of columns. The dimension of the following matrix is 2×3, read "two by three" because it has two rows and three columns.

Matrix	**Dimension**
$\begin{bmatrix} 5 & 6 & -1 \\ 4 & 3 & 1 \end{bmatrix}$	2×3 2 rows by 3 columns

The entries in an augmented matrix for a system of linear equations consist of the coefficients and constants in the equations. To form the augmented matrix for a system, we first align the similar variables on the left side of each equation and the constants on the right side. Then we form each row in the matrix from the coefficients and the constant of the corresponding equation. A ______________ should be written in any position that corresponds to a missing variable in an equation.

System of Linear Equations	**Augmented Matrix**
$\begin{cases} 4x + y = 3 \\ 3x - 2y = 16 \end{cases}$	$\left[\begin{array}{cc\|c} 4 & 1 & 3 \\ 3 & -2 & 16 \end{array}\right]$

Write an augmented matrix for each system of linear equations.

1. $\begin{cases} 3x - 5y = 9 \\ 2x + 7y = 12 \end{cases}$

2. $\begin{cases} 5 - y = 0 \\ 6x + 2y + 7 = 0 \end{cases}$

Write a system of linear equations that is represented by each augmented matrix.

3. $\left[\begin{array}{cc|c} 2 & -1 & 9 \\ -3 & 7 & 10 \end{array}\right]$

4. $\left[\begin{array}{cc|c} 1 & 0 & 3 \\ 0 & 1 & -2 \end{array}\right]$

5. Enter the matrix from problem 3 into your calculator. See Technology Perspective 12.1.1. We will follow up on this in the next section where we will use the calculator to solve systems of linear equations.

6. Give the solution of the system of linear equations represented by $\left[\begin{array}{cc|c} 1 & 0 & 3 \\ 0 & 1 & -2 \end{array}\right]$.

Transformations Resulting in Equivalent Systems

1. Any two equations in a system may be interchanged.
2. Both sides of any equation in a system may be multiplied by a nonzero constant.
3. Any equation in a system may be replaced by the sum of itself and a constant multiple of another equation in the system.

Elementary Row Operations on Augmented Matrices

1. Any two rows in the matrix may be interchanged.
2. Any row in the matrix may be multiplied by a nonzero constant.
3. Any row in the matrix may be replaced by the sum of itself and a constant multiple of another row.

7. Complete the steps to solve the following system of linear equations.

$$\begin{Bmatrix} 2x + y = 1 \\ x - y = 5 \end{Bmatrix} \qquad \left[\begin{array}{cc|c} 2 & 1 & 1 \\ 1 & -1 & 5 \end{array}\right]$$

$$\xrightarrow{r_1 \leftrightarrow r_2} \left[\begin{array}{cc|c} \underline{\quad} & \underline{\quad} & \underline{\quad} \\ \underline{\quad} & \underline{\quad} & \underline{\quad} \end{array}\right]$$

$$\xrightarrow{r_2' = r_2 - 2r_1} \left[\begin{array}{cc|c} 1 & -1 & 5 \\ \underline{\quad} & \underline{\quad} & \underline{\quad} \end{array}\right]$$

$$\xrightarrow{r_2' = \frac{r_2}{3}} \left[\begin{array}{cc|c} 1 & -1 & 5 \\ \underline{\quad} & \underline{\quad} & \underline{\quad} \end{array}\right]$$

$$\xrightarrow{r_1' = r_1 + r_2} \left[\begin{array}{cc|c} \underline{\quad} & \underline{\quad} & \underline{\quad} \\ 0 & 1 & -3 \end{array}\right]$$

$x + 0y =$ ____

$0x + y =$ ____ **Solution:** __________

8. Label the row operations used to solve the following system of linear equations. Then write the solution for this system.

$$\begin{Bmatrix} 3x+12y=9 \\ 4x-3y=-26 \end{Bmatrix} \qquad \left[\begin{array}{cc|c} 3 & 12 & 9 \\ 4 & -3 & -26 \end{array}\right]$$

$$\xrightarrow{r_1' = \frac{r_1}{}} \left[\begin{array}{cc|c} 1 & 4 & 3 \\ 4 & -3 & -26 \end{array}\right]$$

$$\xrightarrow{r_2' = r_2 + \, r_1} \left[\begin{array}{cc|c} 1 & 4 & 3 \\ 0 & -19 & -38 \end{array}\right]$$

$$\xrightarrow{r_2' = \frac{r_2}{}} \left[\begin{array}{cc|c} 1 & 4 & 3 \\ 0 & 1 & 2 \end{array}\right]$$

$$\xrightarrow{r_1' = r_1 + \, r_2} \left[\begin{array}{cc|c} 1 & 0 & -5 \\ 0 & 1 & 2 \end{array}\right]$$

$$x + 0y = -5$$

$$0x + y = 2$$

Solution: ____________

9. Use an augmented matrix and elementary row operations to solve $\begin{Bmatrix} x-2y=-7 \\ 2x+3y=0 \end{Bmatrix}$.

10. The system $\begin{cases} x+2y=3 \\ 3x+6y=3 \end{cases}$ is an inconsistent system. Use an augmented matrix and elementary row operations to solve this system. Write the system of equations for the final matrix. Note the contradiction that occurs during the solution process.

11. The system $\begin{cases} x-2y=-4 \\ -2x+4y=8 \end{cases}$ is a consistent system of dependent equations. Use an augmented matrix and elementary row operations to solve this system. Write a system of equations for the final matrix. Give both the general solution and two particular solutions. Note the identity that occurs during the solution process.

12. The sum of twice one number and a second number is 14. The first number plus three times the second number is 22. What are the two numbers?

(a) Write a system of algebraic equations using the variables x and y and solve the system using augmented matrices and elementary row operations.

(b) Write a sentence that answers the problem.

13. 100 L of a 37% disinfectant solution is prepared by mixing a 45% solution and a 25% solution? How much of each of these solutions are used to make this mixture?

(a) Let x = number of liters of the ________ solution

Let y = number of liters of the ________ solution

(b) Word Equations

______________________ + ______________________ = total solution in mixture

______________________ + ______________________ = total disinfectant in mixture

(c) Algebraic Equations
Write a system of algebraic equations using the variables x and y.

(d) Solve the system using augmented matrices and elementary row operations.

(e) Write a sentence that answers this question. Is this answer reasonable?

12.2 Lecture Guide: Systems of Linear Equations in Three Variables

Objective 1: ***Solve a system of three linear equations in three variables.***

1. Determine whether $(2,-5,6)$ is a solution of $\begin{cases} 2x+y+2z=11 \\ 3x+2y+2z=8 \\ x+4y+3z=0 \end{cases}$

Types of Solution Sets for Linear Systems with Three Equations

The linear system $\begin{cases} A_1x+B_1y+C_1z=D_1 \\ A_2x+B_2y+C_2z=D_2 \\ A_3x+B_3y+C_3z=D_3 \end{cases}$ can have

One Solution

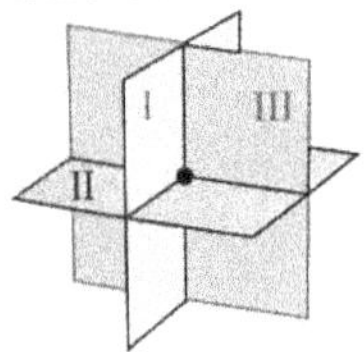

The planes intersect in a single point P; the system is consistent and the equations are independent.

No Solution

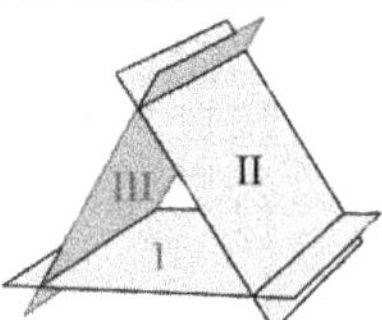

The planes have no point in common; the system is inconsistent.

An Infinite Number of Solutions

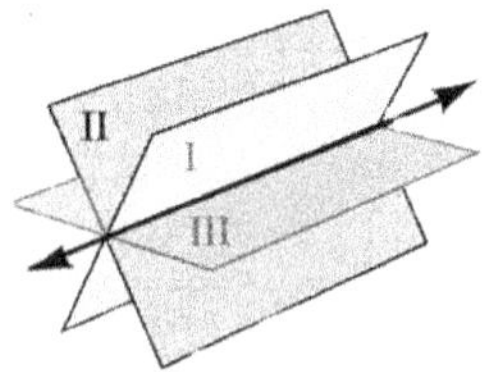

The planes intersect along a line and thus have an infinite number of common points; the system is consistent and the equations are dependent.

Strategy for Solving a 3×3 System of Linear Equations*

Step 1. Write each equation in the general form $Ax + By + Cz = D$.

Step 2. Select one pair of equations and use the substitution or the addition method to eliminate one of the variables.

Step 3. Repeat Step 2 with another pair of equations. Be sure to eliminate the *same* variable as in Step 2.

Step 4. Eliminate another variable from the pair of equations produced in Steps 2 and 3 and solve this system of equations.

Step 5. Back-substitute the values from Step 4 into one of the original equations to solve for the third variable.

Step 6. Does this solution check in all three of the original equations?

*If a contradiction is obtained in any of these steps, the system is inconsistent and has no solution. If an identity is obtained in any step, the system is either dependent with infinitely many solutions or inconsistent with no solution.

2. Solve the System of Linear Equations: $\begin{Bmatrix} x+y+z=2 \\ x-2y-z=2 \\ 4x+3y+2z=4 \end{Bmatrix}$

(a) Produce a 2×2 system of equations.

(b) Produce an equation with only one variable.

(c) Back-substitute to determine the values of the remaining variables.

(d) Write the answer as an ordered triple.

Properties of the Reduced Echelon Form of a Matrix

1. The first nonzero entry in a row is a 1. All other entries in the column containing the leading 1 are zeros.
2. All nonzero rows are above any rows containing only zeros.
3. The first nonzero entry in a row is to the left of the first nonzero entry in the following row.

Transforming an Augmented Matrix into Reduced Echelon Form

Step 1. $\begin{bmatrix} 1 & \cdots \\ 0 & \\ 0 & \\ \vdots & \vdots \\ 0 & \cdots \end{bmatrix}$ **Transform the first column** into this form by using the elementary row operations to
(a) produce a 1 in the top position and
(b) use the 1 in row 1 to produce zeros in the other positions of column 1.

Step 2. $\begin{bmatrix} 1 & 0 & \cdots \\ 0 & 1 & \\ 0 & 0 & \\ \vdots & & \vdots \\ 0 & 0 & \cdots \end{bmatrix}$ **Transform the next column**, if possible, into this form by using the elementary operations to
(a) produce a 1 in the next row and
(b) use the 1 in this row to produce zeros in the other positions of this column.

If it is not possible to produce a 1 in the next row, proceed to next column.

Step 3. Repeat Step 2 column by column, always producing the 1 in the next row, until you arrive at the reduced form.

3. Write an augmented matrix for the system of linear equations.

$$\begin{Bmatrix} 2x+3y-2z=-8 \\ x-y+2z=25 \\ 4x+6y-z=-7 \end{Bmatrix}$$

4. Write a system of linear equations in x, y, and z that is represented by the augmented matrix.

$$\left[\begin{array}{ccc|c} 5 & 6 & -4 & -8 \\ 2 & 4 & -1 & 1 \\ 1 & 1 & -3 & 0 \end{array}\right]$$

5. Complete the steps to solve the following system of linear equations.

$$\begin{Bmatrix} x+3y+2z=1 \\ 2x+y-z=2 \\ x+y+z=2 \end{Bmatrix} \qquad \left[\begin{array}{ccc|c} 1 & 3 & 2 & 1 \\ 2 & 1 & -1 & 2 \\ 1 & 1 & 1 & 2 \end{array}\right]$$

$$\xrightarrow[r_3'=r_3-r_1]{r_2'=r_2-2r_1} \left[\begin{array}{ccc|c} 1 & 3 & 2 & 1 \\ \_\_ & \_\_ & \_\_ & \_\_ \\ \_\_ & \_\_ & \_\_ & \_\_ \end{array}\right]$$

$$\xrightarrow{r_2'=\frac{r_2}{-5}} \left[\begin{array}{ccc|c} 1 & 3 & 2 & 1 \\ \_\_ & \_\_ & \_\_ & \_\_ \\ 0 & -2 & -1 & 1 \end{array}\right]$$

$$\xrightarrow[r_3'=r_3+2r_2]{r_1'=r_1-3r_2} \left[\begin{array}{ccc|c} \_\_ & \_\_ & \_\_ & \_\_ \\ 0 & 1 & 1 & 0 \\ \_\_ & \_\_ & \_\_ & \_\_ \end{array}\right]$$

$$\xrightarrow[r_2'=r_2-r_3]{r_1'=r_1+r_3} \left[\begin{array}{ccc|c} \_\_ & \_\_ & \_\_ & \_\_ \\ \_\_ & \_\_ & \_\_ & \_\_ \\ 0 & 0 & 1 & 1 \end{array}\right]$$

$x+0y+0z=$ ____

$0x+y+0z=$ ____

$0x+0y+z=$ ____ **Solution:** ________

6. Label the row operations used to solve the following system of linear equations. Then write the solution for this system.

$$\begin{cases} x-2y-3z=-6 \\ 3x-5y-z=4 \\ 2x+y+2z=2 \end{cases} \qquad \left[\begin{array}{ccc|c} 1 & -2 & -3 & -6 \\ 3 & -5 & -1 & 4 \\ 2 & 1 & 2 & 2 \end{array}\right]$$

$$\xrightarrow[r_3' = r_3 + \quad r_1]{r_2' = r_2 + \quad r_1} \left[\begin{array}{ccc|c} 1 & -2 & -3 & -6 \\ 0 & 1 & 8 & 22 \\ 0 & 5 & 8 & 14 \end{array}\right]$$

$$\xrightarrow[r_3' = r_3 + \quad r_2]{r_1' = r_1 + \quad r_2} \left[\begin{array}{ccc|c} 1 & 0 & 13 & 38 \\ 0 & 1 & 8 & 22 \\ 0 & 0 & -32 & -96 \end{array}\right]$$

$$\xrightarrow{r_3' = \frac{r_3}{\quad}} \left[\begin{array}{ccc|c} 1 & 0 & 13 & 38 \\ 0 & 1 & 8 & 22 \\ 0 & 0 & 1 & 3 \end{array}\right]$$

$$\xrightarrow[r_2' = r_2 + \quad r_3]{r_1' = r_1 + \quad r_3} \left[\begin{array}{ccc|c} 1 & 0 & 0 & -1 \\ 0 & 1 & 0 & -2 \\ 0 & 0 & 1 & 3 \end{array}\right]$$

$$x+0y+0z=-1$$
$$0x+y+0z=-2$$
$$0x+0y+z=3$$

Solution: ____________

Enter the matrix associated with each system into your calculator and use the **rref** feature to solve the system. Give the reduced form of each matrix and the solution of each system.
See Technology Perspective 12.2.1.

7. $\begin{cases} x-2y+3z=11 \\ 4x+2y-3z=4 \\ 3x+3y-z=4 \end{cases}$

Reduced form:

Solution: ____________

Does this solution check in all three equations?

8. $\begin{cases} 2x+3y-3z=1 \\ x-y+2z=7 \\ 3x+2y+z=6 \end{cases}$

Reduced form:

Solution: ____________

Does this solution check in all three equations?

Forms That Indicate a Dependent or Inconsistent *n x n* System of Equations

Let A be the augmented matrix of an n x n system of equations.

1. If the reduced form of A has a row of the form $[0\ 0\ 0\ \cdots\ 0\ k]$, where $k \neq 0$, then the system is inconsistent and has no solution.
2. If the system is consistent and the reduced form of A has a row of the form $[0\ 0\ 0\ \cdots\ 0\ 0]$ (all zeros), then the system is dependent and has infinitely many solutions.

9. The following system is an inconsistent system. Note the contradiction that occurs in the reduced form of the matrix. Enter the matrix associated with the system into your calculator and use the **rref** feature to solve the system. Give the reduced form of the matrix and the solution of the system.

$$\begin{cases} x+y-z=7 \\ x-y+z=5 \\ 3x+y-z=-1 \end{cases}$$

Reduced form:

Solution: ______________

10. The following system is a consistent system of dependent equations. Note the identity that occurs in the reduced form of the matrix. Enter the matrix associated with the system into your calculator and use the **rref** feature to solve the system. Give the reduced form of the matrix and the solution of the system. Give both the general solution and two particular solutions.

$$\begin{cases} 3x-4y-z=6 \\ 2x-y+z=-1 \\ 4x-7y-3z=13 \end{cases}$$

Reduced form:

General Solution: ______________________

Particular Solution #1: ___________

Particular Solution #2: ___________

11. The sum of three numbers is 105. The third number is eleven less than ten times the second number. Two times the first number is 7 more than three times the second number. What are the numbers?

12.3 Lecture Guide: Horizontal and Vertical Translations of the Graphs of Functions

Objective 1: Analyze and use horizontal and vertical translations.

Vertical Shifts

For a function $y_1 = f(x)$ and a positive real number c:

Algebraically	**Graphically**	**Numerically**
$y_2 = f(x) + c$	To obtain the graph of $y_2 = f(x) + c$, shift the graph of $y_1 = f(x)$ ______________ c units.	For the same x value, $y_2 = y_1 + c$.
$y_3 = f(x) - c$	To obtain the graph of $y_3 = f(x) - c$, shift the graph of $y_1 = f(x)$ ______________ c units.	For the same x value, $y_3 = y_1 - c$.

1. Use your graphing calculator to compare $f(x) = x^2$, $f(x) = x^2 - 3$, and $f(x) = x^2 + 2$ using a graph and a table.
Describe your results:

2. Use the graphs of $y_1 = f_1(x)$ and $y_2 = f_2(x)$ to write an equation for y_2 in terms of $f_1(x)$.

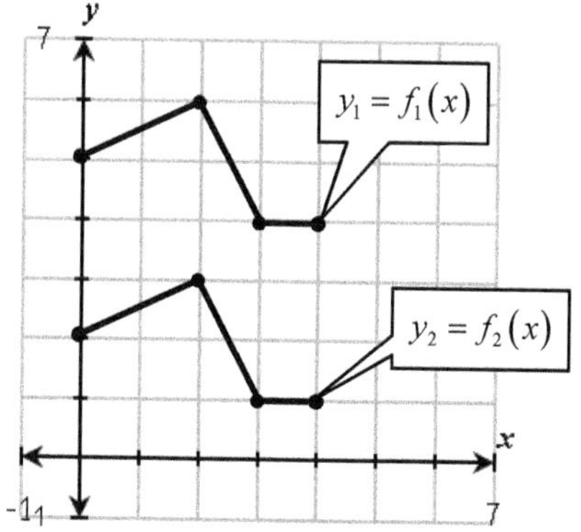

Equation:

3. The graph of $y_1 = f_1(x)$ is shown below. Sketch the graph of $y_2 = f_1(x) + 2$ on the same grid.

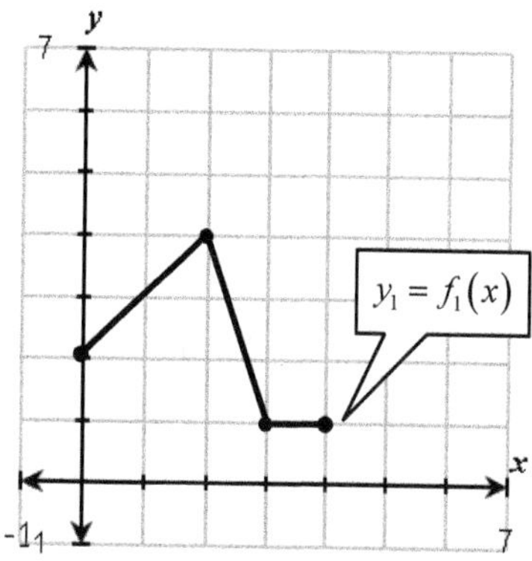

4. Use the table of values for $y_1 = f_1(x)$ and $y_2 = f_2(x)$ to write an equation for y_2 in terms of $y_1 = f_1(x)$.

x	$y_1 = f_1(x)$	$y_2 = f_2(x)$
−3	15	11
−2	2	−2
−1	3	−1
0	3	−1
1	4	0
2	0	−4
3	−1	−5

Equation:

5. The table of values for $y_1 = f_1(x)$ is shown below. Complete the table for $y_2 = f_1(x) + 5$.

x	$y_1 = f_1(x)$	$y_2 = f_1(x) + 5$
−3	6	
−2	7	
−1	−2	
0	8	
1	0	
2	1	
3	−5	

Horizontal Shifts

For a function $y_1 = f(x)$ and a positive real number c:

Algebraically	**Graphically**	**Numerically**
$y_2 = f(x+c)$	To obtain the graph of $y_2 = f(x+c)$, shift the graph of $y_1 = f(x)$ to the ________________ c units.	For the same y-value in y_1 and y_2, the x-value is c units less for y_2 than for y_1.
$y_3 = f(x-c)$	To obtain the graph of $y_3 = f(x-c)$, shift the graph of $y_1 = f(x)$ to the ________________ c units.	For the same y-value in y_1 and y_2, the x-value is c units more for y_2 than for y_1.

6. Use your graphing calculator to compare $f(x) = |x|$, $f(x) = |x-2|$, and $f(x) = |x+3|$ using a graph and a table.

Describe your results:

7. Use the graphs of $y_1 = f_1(x)$ and $y_2 = f_2(x)$ to write an equation for y_2 in terms of $f_1(x)$.

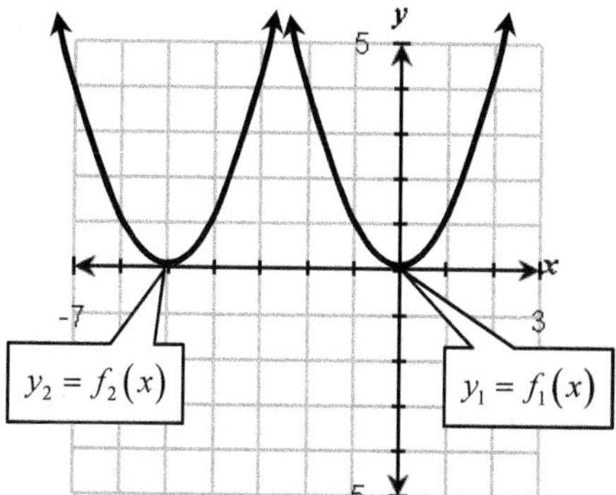

Equation:

8. The graph of $y_1 = f_1(x)$ is shown below. Sketch the graph of $y_2 = f_1(x+2)$ on the same grid.

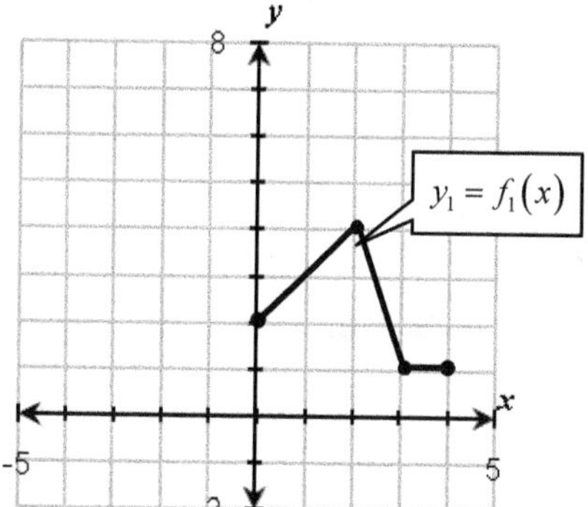

9. Use the table of values for $y_1 = f_1(x)$ and $y_2 = f_2(x)$ to write an equation for y_2 in terms of $y_1 = f_1(x)$.

x	y_1	y_2
−3	5	0
−2	0	−3
−1	−3	−4
0	−4	−3
1	−3	0
2	0	5
3	5	12

Equation:

10. The table of values for $y_1 = f_1(x)$ is shown below. Complete as much as possible of this table for $y_2 = f_1(x-1)$.

x	$y_1 = f_1(x)$	$y_2 = f_1(x-1)$
−3	5	
−2	7	
−1	11	
0	12	
1	10	
2	8	
3	6	

What problems do you encounter when trying to complete this table?

Combining Horizontal and Vertical Shifts:

11. The graph of $y_1 = f_1(x)$ is given. Sketch the graph of y_2 and y_3.

$y_1 = f_1(x)$

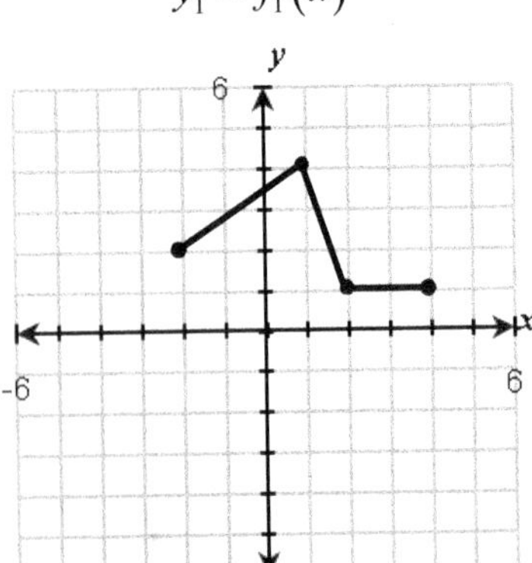

$y_2 = f_1(x+4)$

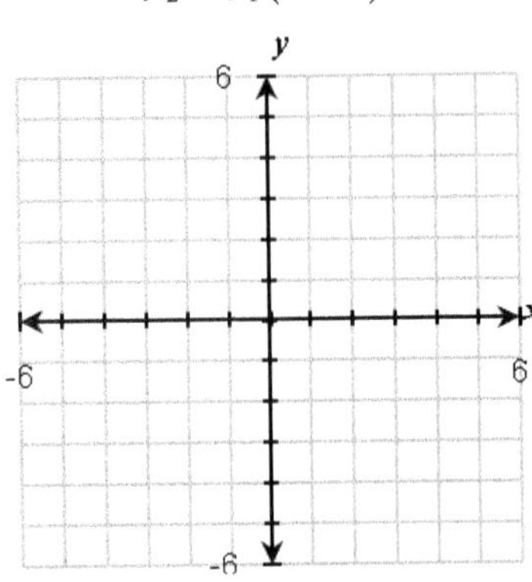

$y_3 = f_1(x+4) - 3$

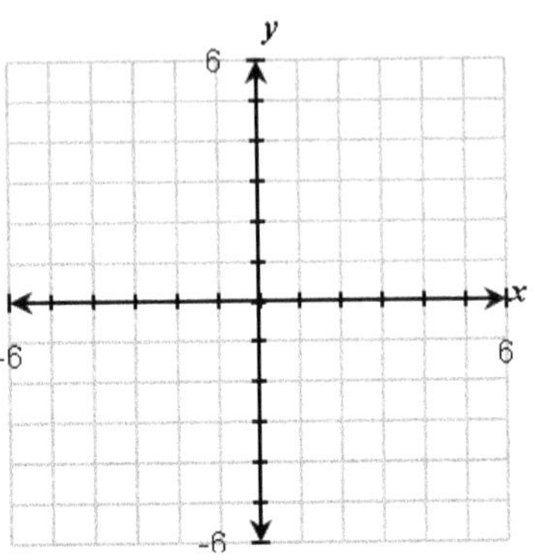

12. The graph of the parabola $f(x) = x^2$ is given. Sketch the graph of y_2 and y_3 and give the location of the vertex of each parabola.

$y_1 = f(x)$

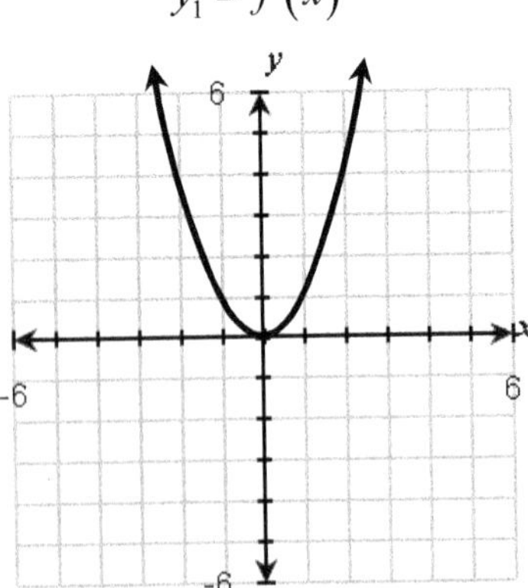

$y_2 = f(x-2)$

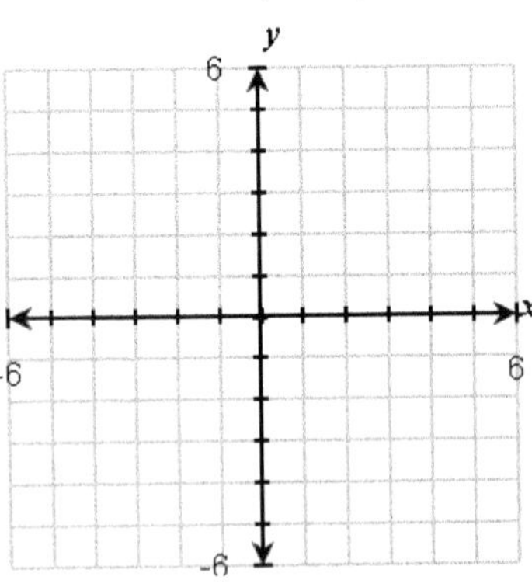

$y_3 = f(x-2) - 3$

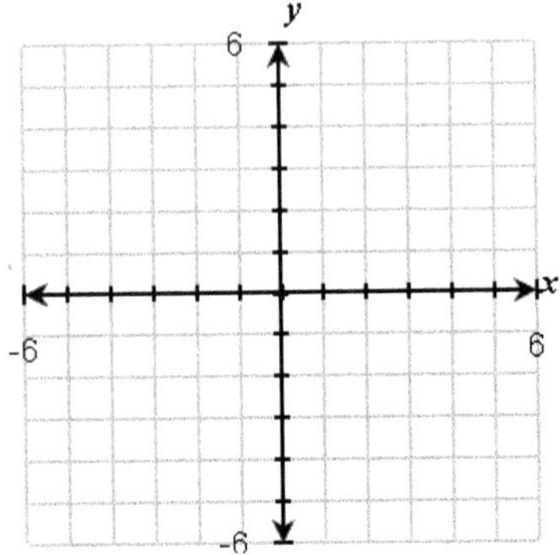

Vertex: __________

Vertex: __________

Vertex: __________

12.4 Lecture Guide: Reflecting, Stretching, and Shrinking Graphs of Functions

Objective 1: Recognize and use the reflection of a graph.

Reflection of $y = f(x)$ Across the x-Axis

Graphically	**Numerically**	**Algebraically**
To obtain the graph of $y = -f(x)$, reflect the graph of $y = f(x)$ across the ______________.	For each value of x, y_2 is the ______________ inverse of y_1. That is, $y_2 = -y_1$.	Original Function: $y_1 = f(x)$ Reflection: $y_2 = -f(x)$

When we reflect a graph across the x-axis, we are creating a mirror image of the graph on the opposite side of the x-axis. Algebraically, this is accomplished by multiplying the expression to be graphed by -1 as illustrated below.

1. Use your graphing calculator to compare $f(x) = x^2$ and $f(x) = -x^2$ using a graph and a table.

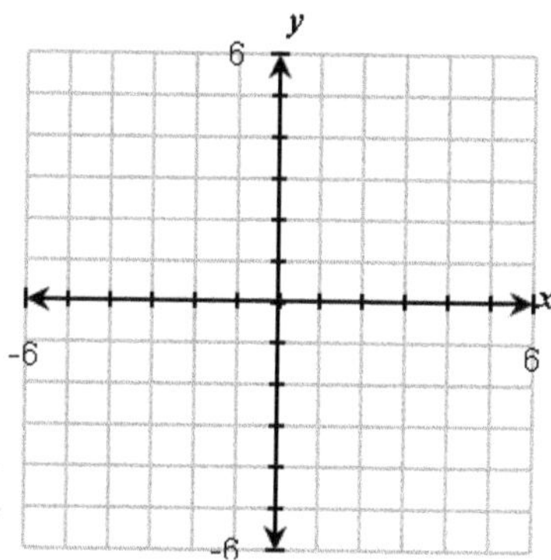

x	$f(x) = x^2$	$f(x) = -x^2$
-2		
-1		
0		
1		
2		

Describe your results:

2. Use your graphing calculator to compare $f(x) = \lvert x \rvert$ and $f(x) = -\lvert x \rvert$ using a graph and a table.

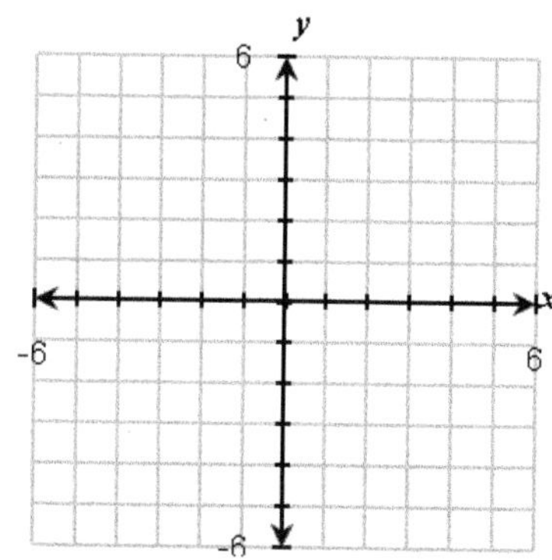

x	$f(x) = \lvert x \rvert$	$f(x) = -\lvert x \rvert$
-2		
-1		
0		
1		
2		

Describe your results:

3. Use the graph of $y_1 = f(x)$ to sketch the graph of $y_2 = -f(x)$.

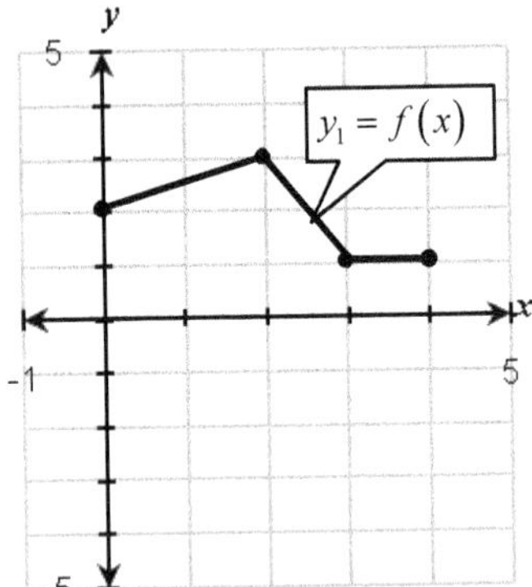

4. Use the graph of $y_1 = f(x)$ to sketch the graph of $y_2 = -f(x)$.

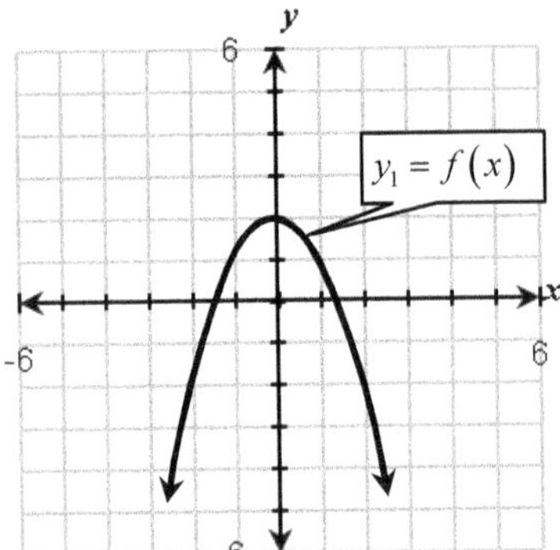

5. If $f(x) = \{(1,3), (2,-4), (5,0), (6,2)\}$ write the ordered pairs for $-f(x)$.

$-f(x) =$

Objective 2: Use stretching and shrinking factors to graph functions.

Stretching and Shrinking Graphs

For a function $y_1 = f(x)$ and a positive real number c:

Algebraically	**Graphically**	**Numerically**
If $c > 1$:		
Original Function: $y_1 = f(x)$	To obtain the graph of $y_2 = cy_1$, vertically ______________ the graph of	For each value of x, $y_2 = cy_1$.
Scaled Function: $y_2 = cf(x)$	$y_1 = f(x)$ by a factor of c.	
If $0 < c < 1$:		
Original Function: $y_1 = f(x)$	To obtain the graph of $y_2 = cy_1$ vertically ______________ the graph of	For each value of x, $y_2 = cy_1$.
Scaled Function: $y_2 = cf(x)$	$y_1 = f(x)$ by a factor of c.	

6. Use your graphing calculator to compare $f(x)=x^2$, $f(x)=2x^2$, and $f(x)=\frac{1}{2}x^2$ using a graph and a table.

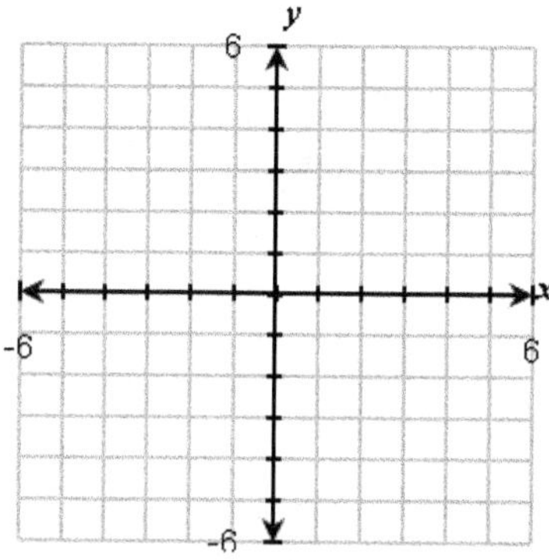

x	$f(x)=x^2$	$f(x)=2x^2$	$f(x)=\frac{1}{2}x^2$
−2			
−1			
0			
1			
2			

Describe your results:

7. Use your graphing calculator to compare $f(x)=|x|$, $f(x)=3|x|$, and $f(x)=\frac{1}{3}|x|$ using a graph and a table.

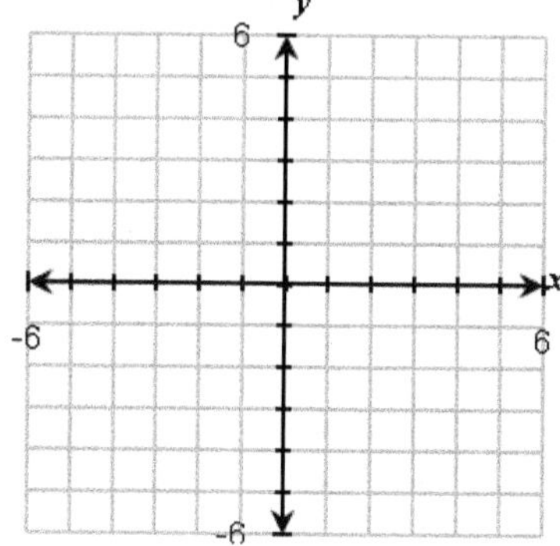

x	$f(x)=\|x\|$	$f(x)=3\|x\|$	$f(x)=\frac{1}{3}\|x\|$
−2			
−1			
0			
1			
2			

Describe your results:

8. Use the graph of $y_1 = f(x)$ to sketch the graph of $y_2 = 2f(x)$ on the same grid.

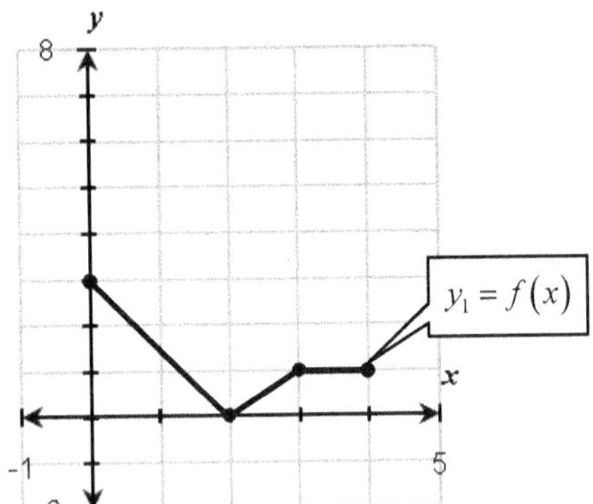

9. Use the graph of $y_1 = f(x)$ to sketch the graph of $y_2 = \frac{1}{2}f(x)$ on the same grid.

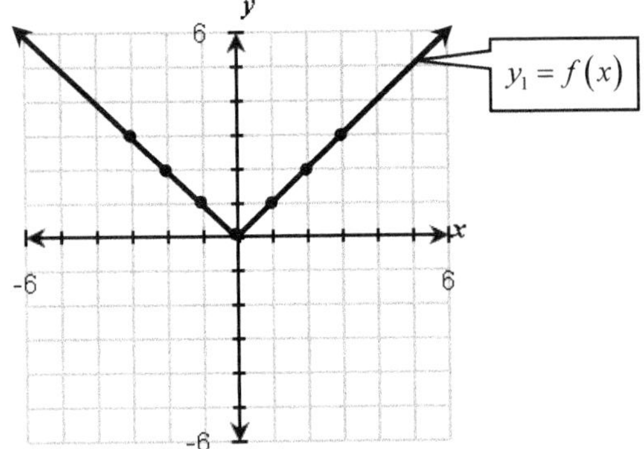

10. Use the table of values for $y_1 = f_1(x)$ and $y_2 = f_2(x)$ to write an equation for y_2 in terms of $y_1 = f_1(x)$.

x	$y_1 = f_1(x)$	$y_2 = f_2(x)$
−3	8	32
−2	5	20
−1	3	12
0	2	8
1	3	12
2	5	20
3	8	32

Equation: $y_2 =$

11. The table of values for $y_1 = f_1(x)$ is shown below. Complete the table for $y_2 = \frac{1}{4}f_1(x)$.

x	$y_1 = f_1(x)$	$y_2 = \frac{1}{4}f_1(x)$
−3	−12	
−2	−8	
−1	−4	
0	0	
1	−4	
2	−8	
3	−12	

Determine the vertex of each parabola using the fact that the vertex of $f(x) = x^2$ is $(0,0)$.

12. $f(x) = \frac{1}{6}x^2$

13. $f(x) = \frac{1}{6}(x-6)^2$

14. $f(x) = (x+3)^2 - 2$

15. $f(x) = -(x+3)^2 + 2$

Combining Stretching, Shrinking, Reflecting, and Translating:
Reflecting, stretching, and shrinking can be used to help us understand more complicated functions by visualizing them as modified versions of basic functions.

16. Use your graphing calculator to assist you in comparing $y_1 = |x|$, $y_2 = |x+1|$, $y_3 = -2|x+1|$, and $y_4 = -2|x+1| - 1$. Describe the steps that were used one at a time to get from y_1 to y_4 and sketch each graph.

$y_1 = |x|$

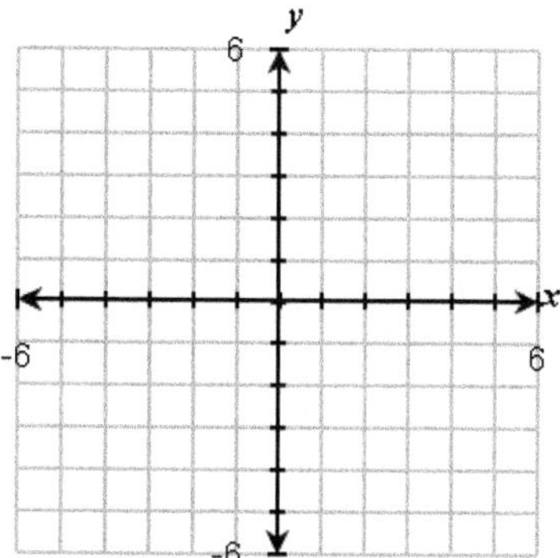

Steps:

$y_2 = |x+1|$

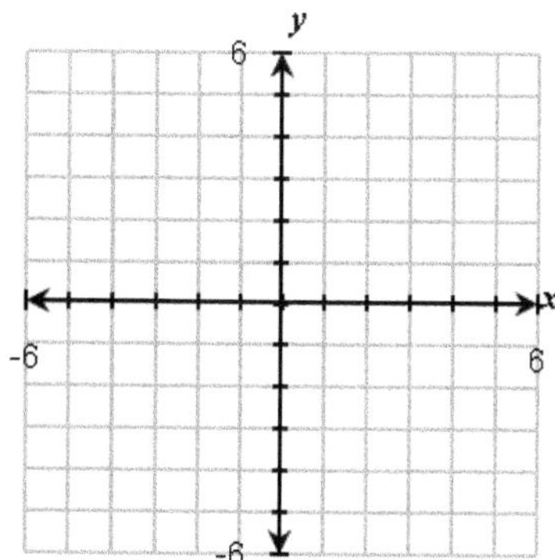

Steps:

$y_3 = -2|x+1|$

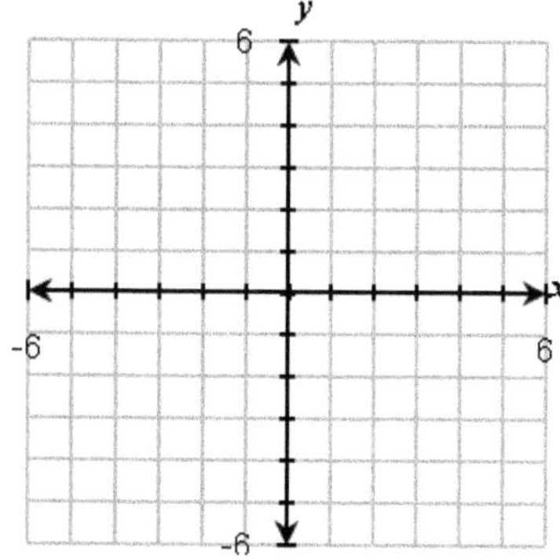

Steps:

$y_4 = -2|x+1| - 1$

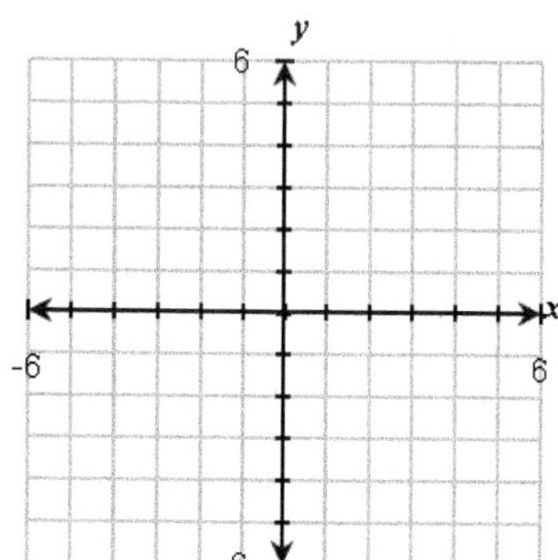

Steps:

17. Use your graphing calculator to assist you in comparing $y_1 = x^2$, $y_2 = (x-3)^2$, $y_3 = -\frac{1}{2}(x-3)^2$, and $y_4 = -\frac{1}{2}(x-3)^2 + 2$. Describe the steps one at a time to get from y_1 to y_4 and sketch each graph.

$y_1 = x^2$

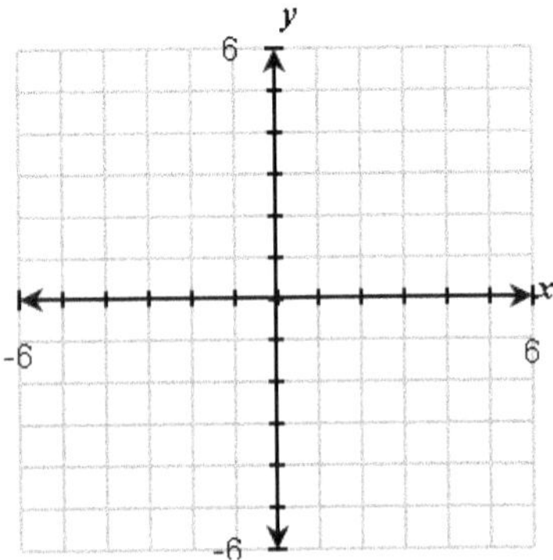

Steps:

$y_2 = (x-3)^2$

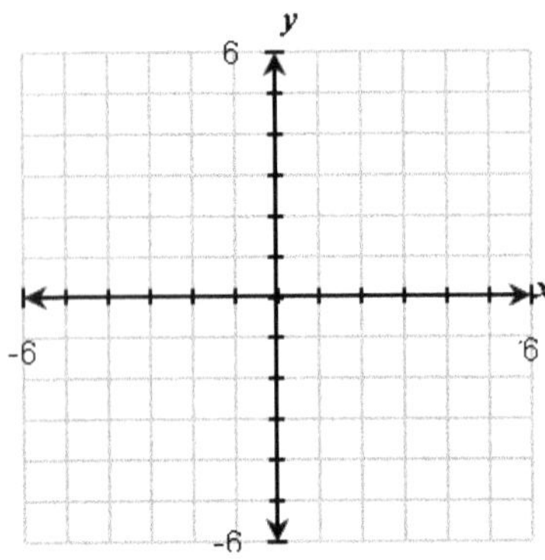

Steps:

$y_3 = -\frac{1}{2}(x-3)^2$

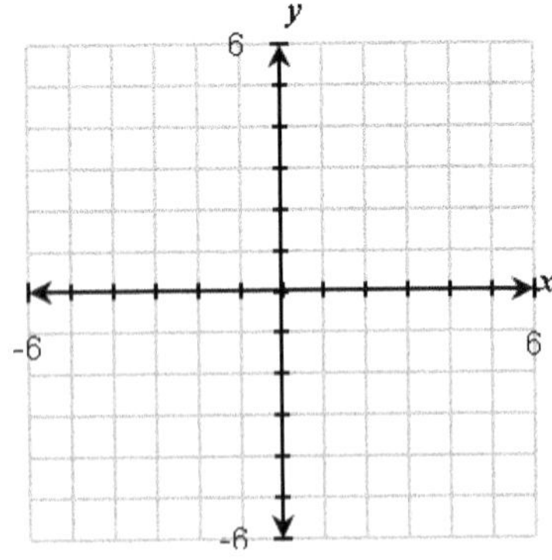

Steps:

$y_4 = -\frac{1}{2}(x-3)^2 + 2$

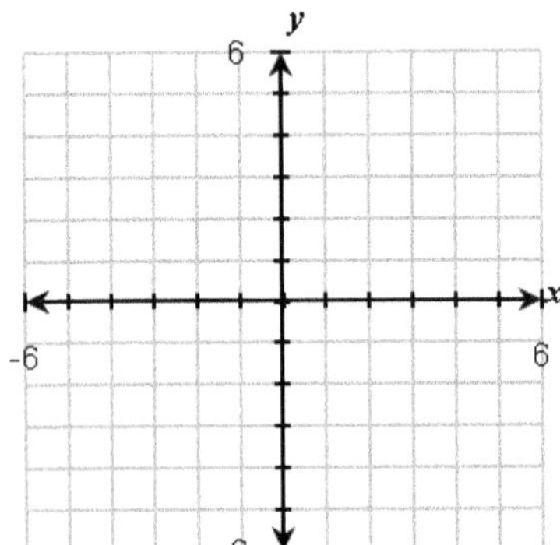

Steps:

Determine the range of each function given that the range of $y = f(x)$ is $(-3, 6]$

18. $y = f(x) - 3$

19. $y = 3f(x)$

20. $y = \frac{1}{3}f(x)$

21. $y = -f(x)$

12.5 Lecture Guide: Algebra of Functions

Objective 1: Add, subtract, multiply, and divide two functions.

Operations of Functions

	Notation	Definition*	Examples for $f(x)=x^2-4$ and $g(x)=x-2$
Sum	$f+g$	$(f+g)(x)=f(x)+g(x)$	$(f+g)(x)=x^2+x-6$ for $\mathbb{R}$
Difference	$f-g$	$(f-g)(x)=f(x)-g(x)$	$(f-g)(x)=x^2-x-2$ for $\mathbb{R}$
Product	$f\cdot g$	$(f\cdot g)(x)=f(x)g(x)$	$(f\cdot g)(x)=x^3-2x^2-4x+8$ for $\mathbb{R}$
Quotient	$\frac{f}{g}$	$\left(\frac{f}{g}\right)(x)=\frac{f(x)}{g(x)}$	$\left(\frac{f}{g}\right)(x)=x+2$ for all reals except 2

*The domain of all of these functions except $\frac{f}{g}$ is the set of values in both the domain of f and the domain of g. For $\frac{f}{g}$, we also must have $g(x)\neq 0$.

Evaluate each expression, given $f(x)=x^2-9$ and $g(x)=2x-3$.

1. $f(2)$

2. $g(2)$

3. $(f+g)(2)$

4. $(f-g)(2)$

5. $(f\cdot g)(2)$

6. $\left(\frac{f}{g}\right)(2)$

Use $f(x)=x^2-9$ and $g(x)=x-3$ to determine each function then find the domain of each function.

7. $(f+g)(x)$

Domain:

8. $(f-g)(x)$

Domain:

9. $(f\cdot g)(x)$

Domain:

10. $\left(\frac{f}{g}\right)(x)$

Domain:

Use the results from problems 7 – 10 to evaluate the following expressions.

11. $(f+g)(-2)$

12. $(f-g)(-2)$

13. $(f\cdot g)(-2)$

14. $\left(\frac{f}{g}\right)(-2)$

Objective 2: Form the composition of two functions.

Composition of Functions

Algebraically	**Verbally**	**Algebraic Example**
$(f \circ g)(x) = f[g(x)]$ The domain of $f \circ g$ is the set of x values from the domain of g for which $g(x)$ is in the domain of f.	$f \circ g$ denotes the composition of function f with function g. $f \circ g$ is read "f composed with g".	For $f(x) = x+1$ and $g(x) = 2x$, $(f \circ g)(x) = f[g(x)]$ $= f(2x)$ $= 2x+1$

Evaluate each expression, given $f(x) = x^2 - 9$ and $g(x) = 2x - 3$.

15. $g(4)$

16. $f(5)$

17. $f(4)$

18. $g(7)$

Use the results above to evaluate each expression.

19. $(f \circ g)(4)$

20. $(g \circ f)(4)$

Use $f(x)=x^2-9$ and $g(x)=2x-3$ to determine each function. Also give the domain of each function.

21. $(f\circ g)(x)$

Domain:

22. $(g\circ f)(x)$

Domain:

Inverse of a Function

Algebraically	**Example**
The functions *f* and f^{-1} are inverses of each other if and only if $(f\circ f^{-1})(x)=x$ for each input value of f^{-1}	$f(x)=3x-1$ and $f^{-1}(x)=\frac{x+1}{3}$ are inverses because: $(f\circ f^{-1})(x)=f\left(\frac{x+1}{3}\right)$ $(f\circ f^{-1})(x)=3\left(\frac{x+1}{3}\right)-1$ $(f\circ f^{-1})(x)=x$
and $(f^{-1}\circ f)(x)=x$ for each input value of *f*	and $(f^{-1}\circ f)(x)=f^{-1}(3x-1)$ $(f^{-1}\circ f)(x)=\frac{(3x-1)+1}{3}$ $(f^{-1}\circ f)(x)=x$

23. Verify that $f(x)=4x+5$ and $f^{-1}(x)=\frac{x-5}{4}$ are inverses of each other by showing that $(f\circ f^{-1})(x)=x$ and $(f^{-1}\circ f)(x)=x$.

12.6 Lecture Guide: Sequences, Series, and Summation Notation

Objective 1: Calculate the terms of arithmetic and geometric sequences.

Arithmetic Sequence

Algebraically	Verbally	Numerical Example
$a_n - a_{n-1} = d$ or $a_n = a_{n-1} + d$	An arithmetic sequence has a ______________ change d from term to term. The graph will form a set of discrete points lying on a straight ______________.	−3, −1, 1, 3, 5, 7

Geometric Sequence

Algebraically	Verbally	Numerical Example
$\frac{a_n}{a_{n-1}} = r$ or $a_n = ra_{n-1}$	A geometric sequence has a constant ______________ r from term to term. The graph will form a set of discrete points lying on an ______________ curve.	$\frac{1}{4}, \frac{1}{2}, 1, 2, 4, 8$

Determine whether each sequence is arithmetic, geometric, both, or neither. If the sequence is arithmetic, write the common difference *d*. If the sequence is geometric, write the common ratio *r*.

1. 3, 12, 48, 192, ...

2. 3, 12, 21, 30, ...

3. 3, − 6, 12, − 24, ...

4. 5, 5, 5, 5, ...

Recursive Definitions

Since a_n can represent any term of the sequence, it is called the general term. The general term of a sequence is sometimes defined in terms of one or more of the ______________ terms. A sequence defined in this manner is said to be defined recursively.

Examples of recursive definitions for problems 1 and 2 above are:

1. $a_n = 4a_{n-1}$ with $a_1 = 3$

2. $a_n = a_{n-1} + 9$ with $a_1 = 3$

Because recursive definitions relate terms to preceding terms, there must be an initial term or condition given so that the sequence can get started. To construct looping structures within their programs computer programmers frequently use this initial or "seed" value.

Formulas for the nth Term of an Arithmetic and a Geometric Sequence

Arithmetic Sequence	Geometric Sequence	
$a_n = a_{n-1} + d$	$a_n = ra_{n-1}$	Start with the recursive formula for a_n and note the pattern that develops. Arithmetic sequences add a common difference for each new term, and geometric sequences multiply by a common ratio for each new term.
a_1	a_1	
$a_2 = a_1 + d$	$a_2 = a_1 r$	
$a_3 = a_2 + d = (a_1 + d) + d = a_1 + 2d$	$a_3 = a_2 r = (a_1 r)r = a_1 r^2$	
$a_4 = a_3 + d = (a_1 + 2d) + d = a_1 + 3d$	$a_4 = a_3 r = (a_1 r^2)r = a_1 r^3$	
$\vdots$	$\vdots$	
$a_n = a_{n-1} + d = a_1 + (n-1)d$	$a_n = a_{n-1} r = a_1 r^{n-1}$	
$\boldsymbol{a_n = a_1 + (n-1)d}$	$\boldsymbol{a_n = a_1 r^{n-1}}$	

5. Determine a_{12} for an arithmetic sequence with $a_1 = 5$ and $d = 3$.

6. Determine a_{12} for a geometric sequence with $a_1 = 5$ and $r = 3$.

7. Find d in an arithmetic sequence with $a_1 = -85$ and $a_{51} = 265$.

8. Find r in a geometric sequence with $a_1 = 4$ and $a_3 = 36$.

Determine whether each sequence is arithmetic or geometric. Then use the appropriate formula to calculate the number of terms in each sequence.

9. 5, 14, 23, …, 131

10. 5, 15, 45, …, 32805

Objective 2: Use summation notation and evaluate the series associated with a finite sequence.

Series

The sum of the terms of a sequence is called a **series**. If $a_1, a_2, \ldots, a_n$ is a finite sequence, then the indicated sum $a_1 + a_2 + \ldots + a_n$ is the series associated with this sequence. For arithmetic and geometric series there are formulas that serve as shortcuts for evaluating the series. A convenient way of denoting a series is to use summation notation, which uses the Greek letter Σ (*sigma*, which corresponds to *S* for "sum") to indicate the summation.

Summation Notation

Algebraically	Verbally	Algebraic Example
$\sum_{i=1}^{n} a_i = a_1 + a_2 + \cdots + a_{n-1} + a_n$	The sum of *a* sub *i* from *i* equals 1 to *i* equals *n*.	$\sum_{i=1}^{5} a_i = a_1 + a_2 + a_3 + a_4 + a_5$

Write the terms of each series and then add these terms. Describe each series as arithmetic or geometric.

11. $\sum_{i=1}^{3} 2i$

12. $\sum_{k=2}^{4} 3^k$

Formulas for Arithmetic and Geometric Series: $S_n = \sum_{i=1}^{n} a_i$

	Algebraically	Algebraic Example
Arithmetic Series:	$S_n = \frac{n}{2}(a_1 + a_n)$	$S_4 = \sum_{i=1}^{4} 5i = 5 + 10 + 15 + 20$
	or	$S_4 = \frac{4}{2}(5+20)$
	$S_n = \frac{n}{2}\left[2a_1 + (n-1)d\right]$	$S_4 = 2(25)$ $S_4 = 50$
Geometric Series:	$S_n = \frac{a_1(1-r^n)}{1-r}$	$S_5 = \sum_{k=1}^{5} 2^k = 2 + 4 + 8 + 16 + 32$
	or	$S_5 = \frac{2(1-2^5)}{1-2}$
	$S_n = \frac{a_1 - ra_n}{1-r}$	$S_5 = \frac{2(-31)}{-1}$ $S_5 = 62$

Use the formulas for arithmetic and geometric series to determine each sum. (Hint: First determine whether the series is arithmetic or geometric.)

13. $\sum_{i=1}^{200} 3i$

14. $\sum_{i=1}^{10} 3^i$

15. S_{25} for the sequence defined by $a_n = 3 + (n-1)5$

16. S_7 for the sequence defined by $a_n = 1028\left(\frac{1}{2}\right)^{n-1}$

Objective 3: Evaluate an infinite geometric series.

Infinite Geometric Series: $S=\sum_{i=1}^{\infty} a_i$

Algebraically	**Algebraic Example**
If $\lvert r\rvert<1$, $S=\dfrac{a_1}{1-r}$	$S=0.333\ldots$ $S=0.3+0.03+0.003+\ldots$ with $a_1=0.3$ and $r=0.1$. $S=\dfrac{a_1}{1-r}$ $S=\dfrac{0.3}{1-0.1}$ $S=\dfrac{0.3}{0.9}$ $S=\dfrac{1}{3}$
If $\lvert r\rvert\geq 1$, this sum does not exist.	

Evaluate each series if it exists.

17. $\sum_{i=1}^{\infty}\left(\frac{2}{3}\right)^i$

18. $\sum_{i=1}^{\infty}(4)^i$

19. Write $0.515151\ldots$ as a fraction.

12.7 Lecture Guide: Conic Sections

Objective 1: Graph the conic sections and write their equations in standard form.

Distance and Midpoint Formulas

If (x_1, y_1) and (x_2, y_2) are two points, then:

Algebraically

Distance Formula: $d = \sqrt{(x_2 - x_1)^2 + (y_2 - y_1)^2}$

Midpoint Formula:

$(x, y) = \left(\frac{x_1 + x_2}{2}, \frac{y_1 + y_2}{2}\right)$

Numerical Example

For (–2, 7) and (4, –1):
Distance between the points:

$d = \sqrt{(4-(-2))^2 + (-1-7)^2}$

$d = \sqrt{36+64} = \sqrt{100}$

$d = 10$

Midpoint between the points:

$(x, y) = \left(\frac{-2+4}{2}, \frac{7+(-1)}{2}\right) = \left(\frac{2}{2}, \frac{6}{2}\right)$

$(x, y) = (1, 3)$

1. Consider the following pair of points: $(4, 3)$ and $(-2, 11)$

(a) Determine the distance between the two points.

(b) Determine the midpoint of the pair of points.

Standard Form of the Equation of a Circle

Algebraically

The equation of a circle with center (h, k) and radius r is $(x-h)^2 + (y-k)^2 = r^2$.

Graphically

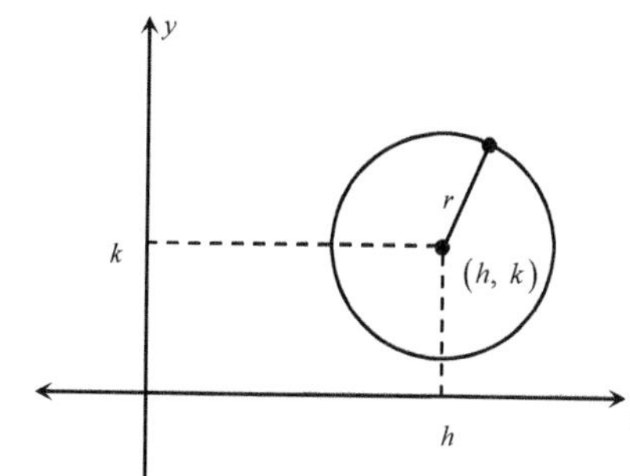

Algebraic Example

The equation of a circle with center (–1, 3) and radius 5 is:

$(x-(-1))^2 + (y-3)^2 = 5^2$

$(x+1)^2 + (y-3)^2 = 25$

2. Give the equation of the circle that has a diameter with endpoints $(4, 3)$ and $(-2, 11)$. Then sketch a graph of the circle. Hint: Use your work from problem 1.

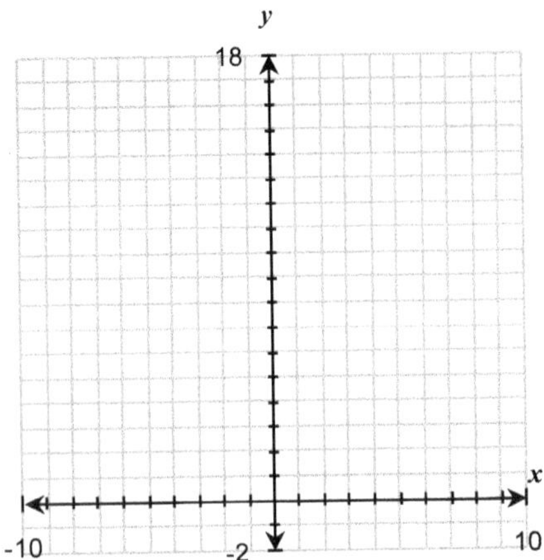

Equations of the Form $x^2 + y^2 = c$

As shown in the figure to the right,

$x^2 + y^2 = 4$ is a circle with center (0,0) and radius _____.

$x^2 + y^2 = 1$ is a circle with center (0,0) and radius _____.

$x^2 + y^2 = \frac{1}{4}$ is a circle with center (0,0) and radius _____.

$x^2 + y^2 = 0$ is the degenerate case of a circle – a single point, (0,0).

$x^2 + y^2 = -1$ has no real solutions and thus no points to graph. Both x^2 and y^2 are nonnegative, so $x^2 + y^2$ must also be nonnegative.

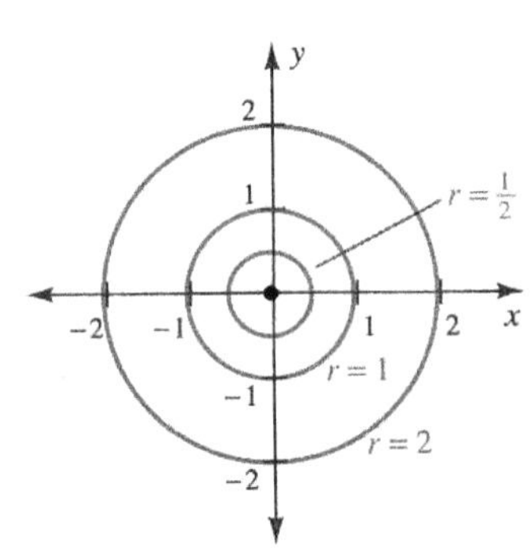

Sketch the graph of each circle.

3. $x^2 + (y-2)^2 = 9$

4. $(x-2)^2 + (y+1)^2 = 4$

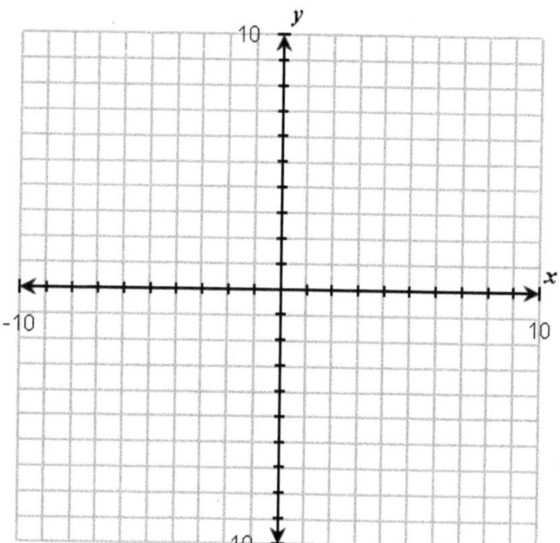

Write in standard form the equation of the circle satisfying the given conditions.

5. Center $(0, 0)$, radius 3

6. Center $(4, -1)$, radius 5

Standard Form of the Equation of an Ellipse

Algebraically

The equation of an ellipse with center (h, k), major axis of length $2a$, and minor axis of length $2b$ is:

Horizontal Major Axis

$$\frac{(x-h)^2}{a^2}+\frac{(y-k)^2}{b^2}=1$$

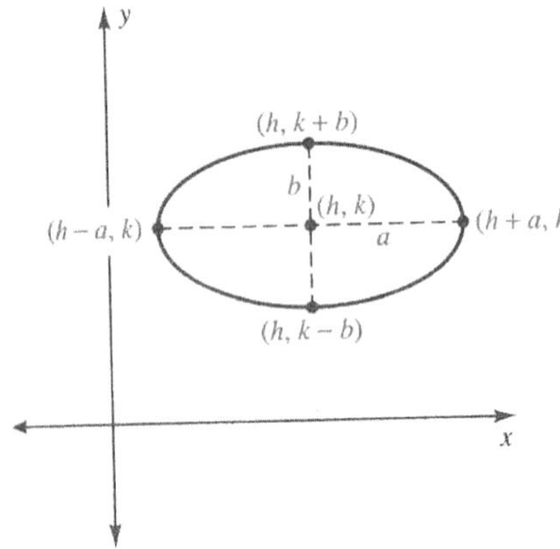

Algebraic Example

$$\frac{(x-2)^2}{49}+\frac{(y+3)^2}{16}=1$$

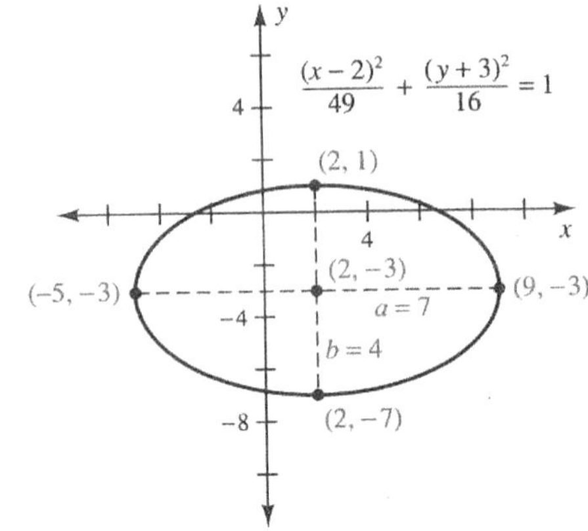

Vertical Major Axis

$$\frac{(x-h)^2}{b^2}+\frac{(y-k)^2}{a^2}=1$$

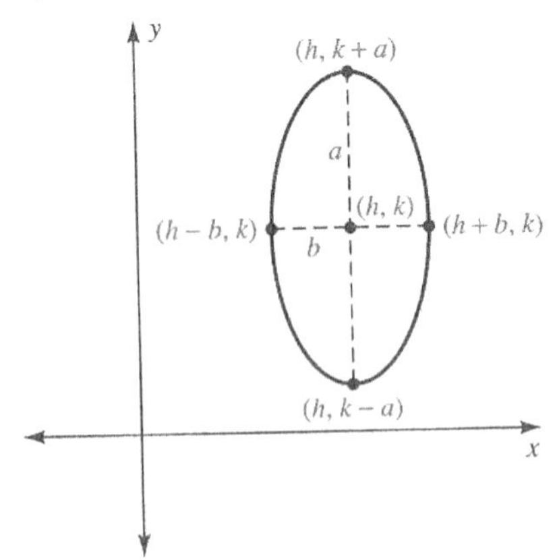

$$\frac{(x+4)^2}{16}+\frac{(y-5)^2}{36}=1$$

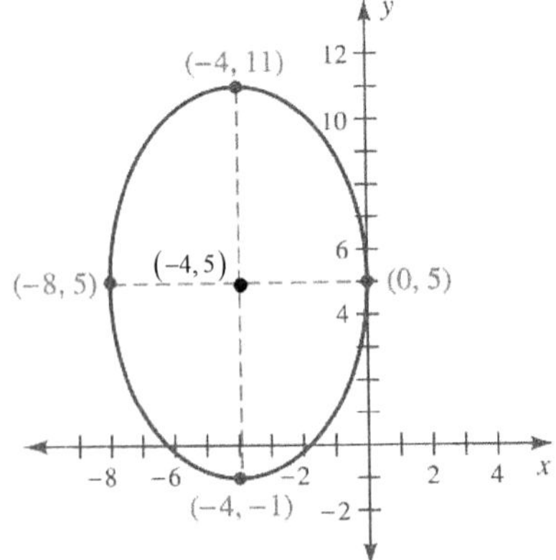

Note: $a > b > 0$.

Sketch the graph of each ellipse.

7. $\frac{(x+2)^2}{49}+\frac{(y-3)^2}{9}=1$

8. $\frac{(x-1)^2}{36}+\frac{(y-4)^2}{4}=1$

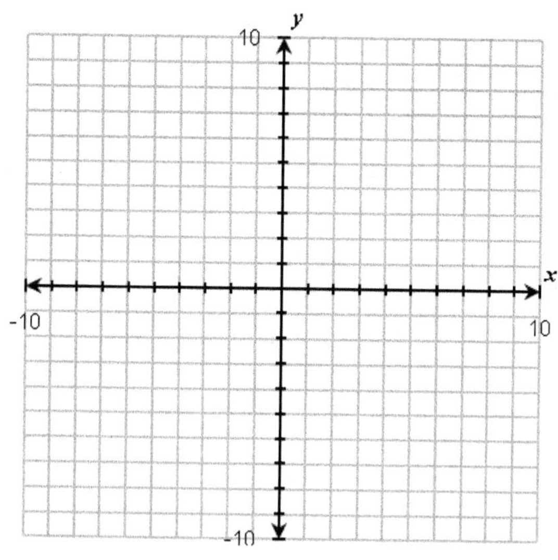

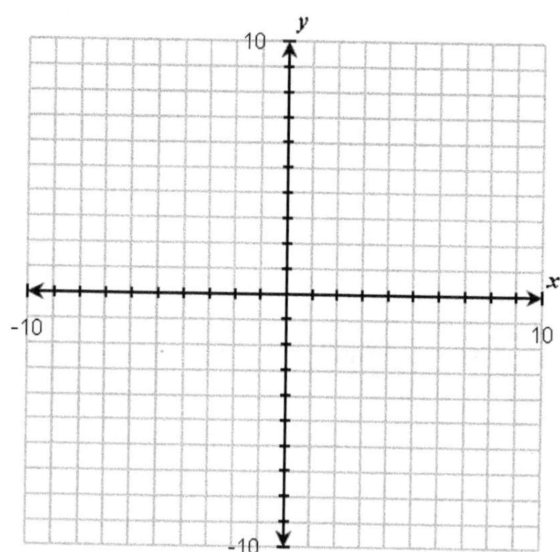

Write in standard form the equation of the ellipse satisfying the given conditions.

9. Center $(0, 0)$, x-intercepts $(-3, 0)$ and $(3, 0)$ and y-intercepts $(0, -7)$ and $(0, 7)$

10. Center $(-3, 4)$, horizontal major axis of length 8, vertical minor axis of length 10

Standard Form of the Equation of a Hyperbola

Algebraically

The equation of a hyperbola with center (h, k) is:

Opening Horizontally

$$\frac{(x-h)^2}{a^2}-\frac{(y-k)^2}{b^2}=1$$

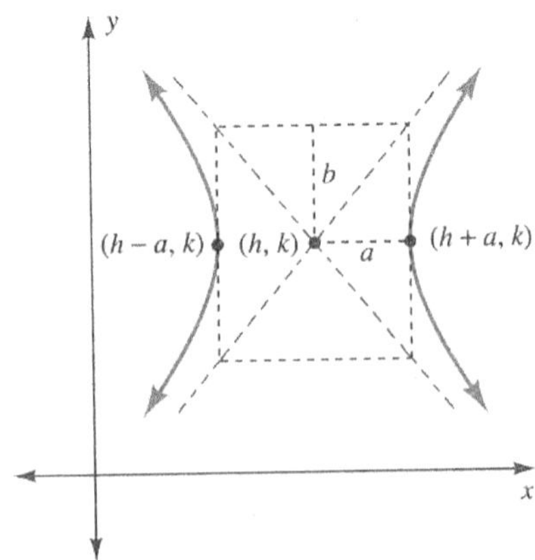

Vertices: $(h-a, k)$ and $(h+a, k)$.
The fundamental rectangle has base $2a$ and height $2b$.

Algebraic Example

$$\frac{(x+1)^2}{16}-\frac{(y-2)^2}{9}=1$$

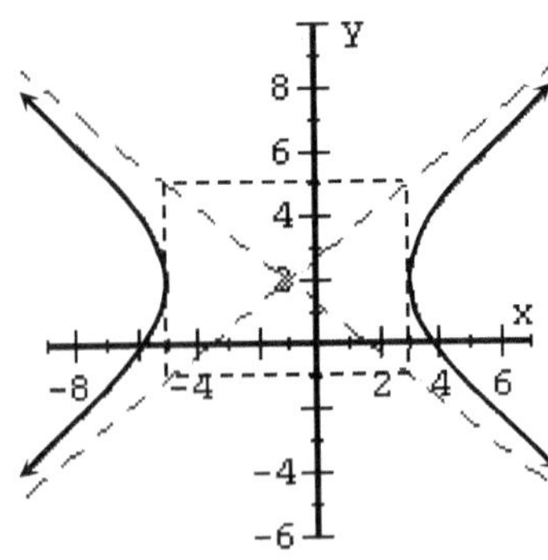

Opening Vertically

$$\frac{(y-k)^2}{a^2}-\frac{(x-h)^2}{b^2}=1$$

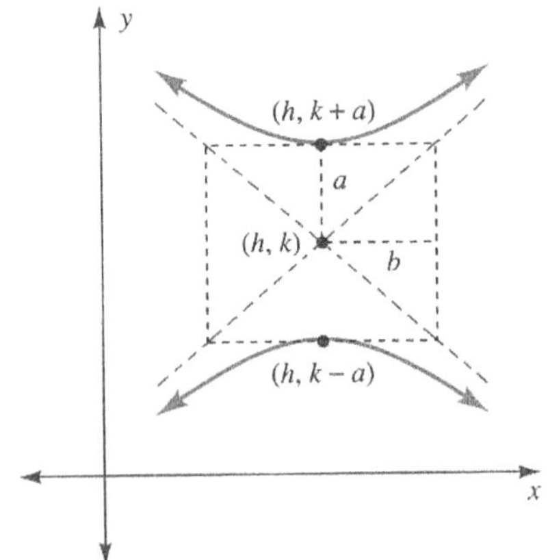

Vertices: $(h, k-a)$ and $(h, k+a)$.
The fundamental rectangle has base $2b$ and height $2a$.

$$\frac{(y-2)^2}{9}-\frac{(x+1)^2}{16}=1$$

Sketch the graph of each hyperbola.

11. $\frac{(x-2)^2}{16}-\frac{(y+1)^2}{25}=1$

12. $\frac{(y+2)^2}{25}-\frac{(x-3)^2}{9}=1$

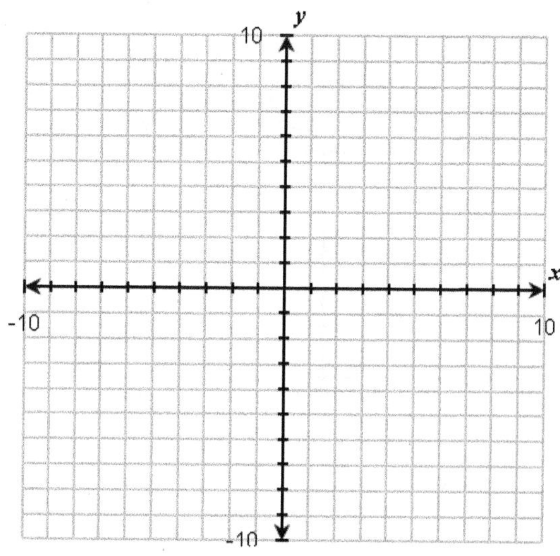

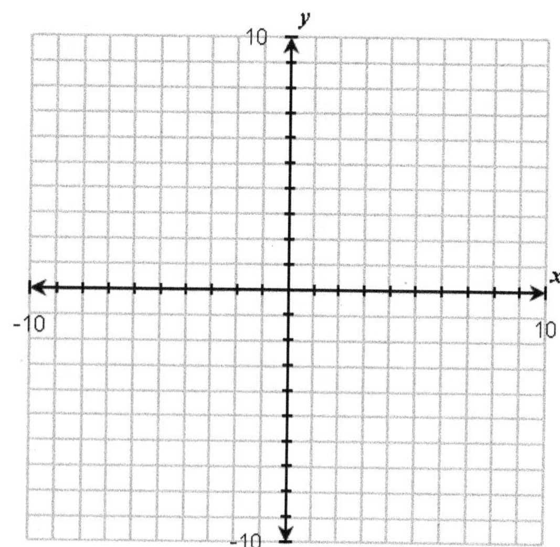

Write in standard form the equation of the hyperbola satisfying the given conditions.

13. Center $(0, 0)$, the hyperbola opens horizontally, and the fundamental rectangle has height 8 and width 14.

14. Vertices $(-4, 0)$ and $(4, 0)$, and the height of the fundamental rectangle is 12.

Parabolas

We have already graphed parabolas in several sections of this book. One way to graph theses parabolas is to use horizontal and vertical shifts of $y = x^2$ as shown in Section 12.3 and reflections as shown in Section 12.4.

Sketch the graph of each parabola.

15. $y = (x-2)^2 - 3$

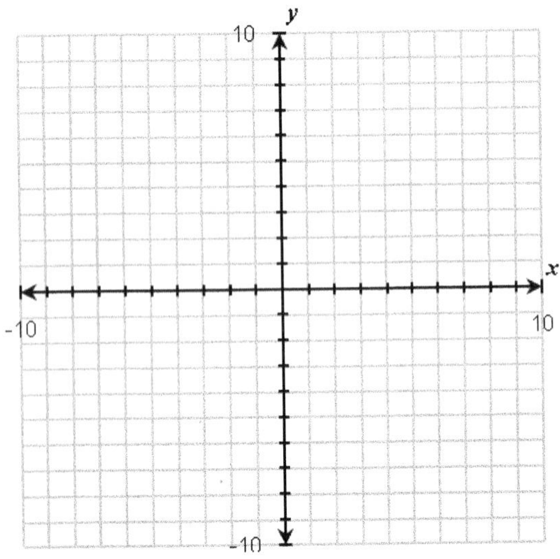

16. $y = -(x+3)^2 + 4$

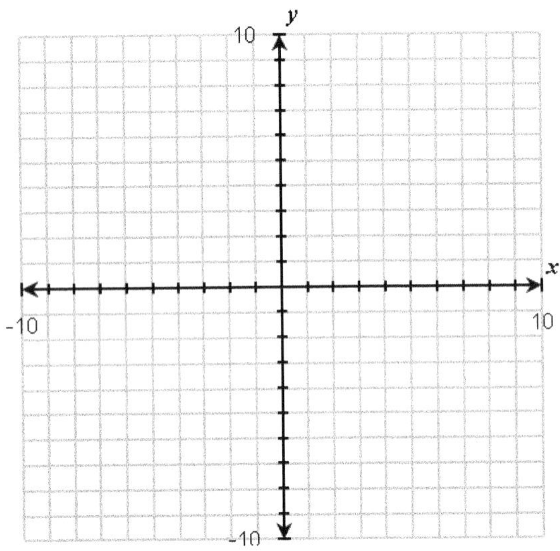

Notes

Notes

Notes

Notes

Notes

Notes